D1103744

The State of Science

*What the Future Holds and
the Scientists Making It Happen*

Marc Zimmer

Prometheus Books
Guilford, Connecticut

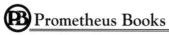
Prometheus Books

An imprint of The Rowman & Littlefield Publishing Group, Inc.
4501 Forbes Boulevard, Suite 200
Lanham, Maryland 20706
www.rowman.com

Distributed by NATIONAL BOOK NETWORK

British Library Cataloguing in Publication Information Available

Library of Congress Cataloging-in-Publication Information

Name: Zimmer, Marc, author.
Title: The state of science : what the future holds and the scientists making it happen / Marc Zimmer.
Description: Guilford, Connecticut : Prometheus Books, [2020] | Includes bibliographical references. | Summary: "New research and innovations in the field of science are leading to life-changing and world-altering discoveries like never before. What does the horizon of science look like? Who are the scientists that are making it happen? And, how are we to introduce these revolutions to a society in which a segment of the population has become more and more skeptical of science? Climate change is the biggest challenge facing our nation, and scientists are working on renewable energy sources, meat alternatives, and carbon dioxide sequestration. At the same time, climate change deniers and the politicization of funding threaten their work. CRISPR (clustered regularly interspaced short palindromic repeats) repurposes bacterial defense systems to edit genes, which can change the way we live but also presents real ethical problems. Optogenetics will help neuroscientists map complicated neural circuitry deep inside the brain, shedding light on treating Alzheimer's and Parkinson's disease. Zimmer also investigates phony science ranging from questionable 'health' products to the fervent anti-vaccination movement. Zimmer introduces readers to the real people making these breakthroughs. Concluding with chapters on the rise of women in STEM fields, the importance of U.S. immigration policies to science, and new, unorthodox ways of DIY science and crowdsource funding, The State of Science shows where science is, where it is heading, and the scientists who are at the forefront of progress"—Provided by publisher.
Identifiers: LCCN 2019054457 (print) | LCCN 2019054458 (ebook) | ISBN 9781633886391 (cloth) | ISBN 9781633886407 (epub)
Subjects: LCSH: Science—History. | Science—Social aspects. | Science—Forecasting.
Classification: LCC Q172 .Z56 2020 (print) | LCC Q172 (ebook) | DDC 500—dc23
LC record available at https://lccn.loc.gov/2019054457
LC ebook record available at https://lccn.loc.gov/2019054458

∞™ The paper used in this publication meets the minimum requirements of American National Standard for Information Sciences Permanence of Paper for Printed Library Materials, ANSI/NISO Z39.48-1992.

Contents

Part Six: Future Science

Part One

Science

Chapter One

The Big Picture

State of the Union address: A yearly speech given by the U.S. president to Congress and the people to tell them about important things that are affecting the country.—*Merriam-Webster Collegiate Dictionary*

Every year the leaders of countries, corporations, and universities present their views of the current conditions of their respective institutions. Given the function of these statements, they are often a top-down view with a biased perspective designed to make their current administrations look good.

There is no such annual report on science. This book is a report on the state of science from a practicing scientist, a view from the trenches. It is not meant to be a top-down view, nor is it meant to make anyone look good or bad.

I have written this book because science is not the same as it was in the first 20–25 years of my career. The way science is done, funded, and disseminated is evolving. At the same time, science is being politicized, and the public's trust in science is being undermined. Yet science remains the engine of our economy and is responsible for our improved well-being. And perhaps now more than ever before, it is at the cusp of altering the most fundamental aspects of our daily lives.

WHAT IS SCIENCE?

This is probably a good point in the book to define "science" and "scientist." The *Merriam-Webster Collegiate Dictionary* defines science as "the state of knowing" as distinguished from ignorance or misunderstanding." However, I much prefer Karl Popper's perspective, according to which a hypothesis has to be inherently disprovable for it to be a scientific theory. It is difficult to imagine an experiment that can disprove the existence of beauty and God. Hence the study of beauty and God is not science. Albert Einstein proposed the existence of gravitational waves. Their existence was provable, although the equipment to do so didn't exist when Einstein came up with his general theory of relativity, and it took close to 100 years to actually prove that they exist (chapter 6). It was disprovable and therefore a scientific theory. Scientists use experiments to test their theories and hypotheses. No map shows the way to prove or disprove a theory; the scientific process is not linear. Neil Gershenfeld, director of MIT's Center for Bits and Atoms, writes that "to find something that's not already on the map, you need to leave the road and wander in the woods besides it." He feels that nonscientists do not recognize that "science appears to be goal-directed only after the fact. While it's unfolding, it is more like a chaotic dance of improvisation, than a victory march."[1] It is also important to realize that uncertainty is an inherent and unremovable component of scientific experimentation. It is not a weakness; it is a strength. Critics of science often disparage scientific results on the basis of included uncertainties. This is a mistake born from a lack of understanding.

Before 1833, people who conducted experiments—who mixed, observed, and synthesized chemicals—were called natural philosophers. Newton, Galen, Galileo, and Mendel were all natural philosophers. In 1833 William Whewell, a professor at Cambridge University, coined the term "scientist" to highlight the fact that these were empirical folks and not philosophers of ideas. Like an artist works with art, a scientist works with science. As an analogy of "artist," the word "scientist" was infused by Whewell with creativity, intuition, and professionalism.

There is a commonly told tale that Whewell derived and popularized the term "scientist" because he needed an alternative to the commonly used phrase "man of science" to describe Mary Somerville in a positive review he was writing about her.[2] If true this would be a great story, and Mary Somerville was worthy of the honor. Her book *On the Connexion of the Physical Sciences* was instrumental in making modern physics into a discipline and was frequently revised and republished, with nine separate editions. In 1879, the first women's college at Oxford University was named in her honor. A main-belt asteroid (5771 Somerville) and a lunar crater in the eastern part of the moon are also named after her, and she graces all Scottish ten-pound notes printed from October 2018 onward. However, according to James Secord, professor of philosophy of science at the University of Cambridge, "Nowhere did Whewell or anyone else in her lifetime ever call Somerville a scientist, nor is it a word, so far as we know, that she ever used herself. By our current understanding of the term, Somerville can certainly be called a scientist, but for her contemporaries she belonged to a higher and more profound category entirely."[3]

No matter whether Mary Somerville was the woman for whom the word "scientist" was made, we can think of scientists as empiricists who use their creativity to prove or disprove theories. Of course scientists are human, as they can discover gloriously complex theories while also misinterpreting and presenting data in wondrously imperfect ways that lead to fantastic new models that are wrong and subsequently disproven. That is the way scientists perform science.

Before going on a trip through the scientific world and trying to judge its impact on our lives, as well as its reputation and health, it is a good idea to take a few steps back and look at the big picture. Because when we talk about science, it is impossible to ignore the political, environmental, economic, and global surroundings. We need to know how science has gotten us here and what challenges it faces in today's world.

SCIENCE AND THE ANTHROPOCENE

Since the arrival of the first *Homo sapiens* 40,000 years ago, a total of 108 billion humans have lived on Earth. A 15th of all these people are alive and kicking right now. Science and technology are the reason the numbers have undergone such robust growth.

Thanks to the development of weapons, protective houses, medicine, fertilizer, and so forth, humans have moved from being an occasional snack for predators to being at the top of the food chain. We have overcome constraints imposed by nature and evolution. There are no predators, other than ourselves, to limit our growth. Humans have become, in the worst sense of the word, the dictators of nature. Thanks to science, we can impose our will on nature and determine our own destinies. We are not doing a great job. Currently we are losing species 1,000 times faster than the natural rate of extinction. Humans represent just 0.01 percent of all living things by mass, yet during our time on Earth we have caused the loss of 83 percent of all wild mammals and half of all plants.[4]

Natural selection has been around for four billion years, since life on Earth began. But humans as a species are no longer governed by it. Not only are we in charge of our own evolution, but we have also changed how nature evolves. Ever since the continents broke apart, delicate ecosystems have been isolated by mountain ranges and oceans. Increased human movement has made the borders between ecosystems more porous. We have both inadvertently and purposefully introduced new species into ecosystems they could never have visited without hitching a ride on our cars, ships, and planes. In some sense, we have created a whole new global pseudo-ecosystem.[5] At the same time, thanks to modern science (CRISPR, gene drives, etc.; chapter 9) we have the ability to control the evolution of other species.

In 2000 Paul Crutzen, an atmospheric chemist and Nobel laureate, started popularizing the concept of the Anthropocene, a new geological age, during which human activity has been the dominant influence on climate and the environment. The International Commission on Stratigraphy, which is in charge of approving and naming subdivisions of geological time, hasn't yet officially approved the use of "Anthropocene" but is working on it. In an important first step, a preliminary Anthropocene working group supported the proposal and suggested the epoch start in the mid-20th century, in part because radioactive debris from the first atomic bomb is a distinct part of the geological record. Artificial radionuclides are just one of many traces we will leave as signatures in the strata of time. Traces of plastics, nitrogen-rich fertilizers, and fossils of domesticated animals and livestock species are just some of the many remnants we will leave behind in the anthropogenic layers of rock.

The International Commission on Stratigraphy will most likely vote on introducing the label for the new epoch in 2021.[6] For now, we are still officially in the Holocene era. Nevertheless, there can be little doubt that we are in an age in which humankind, through science and technology, possesses unprecedented control over its surroundings and nature.

Science, Economy, and Equity

Science and its resultant technologies are also responsible for the well-being of our economies. Our financial systems rely on growth. To be successful in today's global market, the economies of our countries are expected to grow at increasing rates. Games such as The Settlers of Catan, Civilization, Risk, Monopoly, and Minecraft, in which one has to expand one's possessions to win the game, are microcosms of this need. There isn't much room today to grow our agriculture or industries, and fortunately the days of colonization are over, so finding new science-based technologies is one of the few ways we can increase our economies. This places pressure on scientists to produce, which can result in corners being cut and science moving too fast for the global community to establish ethical and safe boundaries.

Scientific discoveries have not only enabled the expansion of the human population but are also the key to expanding economies in the postindustrial world. To keep the financial systems growing, such findings need to come faster and faster. At the same time, scientific discoveries will have to curtail the environmental and ecological degradation associated with growth. These expectations are unrealistic and a sign of a broken system; science cannot deliver all this. Expecting science to rescue us from the consequences of our overconsumption is just as dangerous as failing to acknowledge its great potential. In *Falter: Has the Human Game Begun to Play Itself Out?*, Bill McKibben argues that we have to move from a growth economy to a mature economy.[7] To illustrate his thinking, he uses a human analogy: as teens we are expected to grow and would be taken to the doctor if we didn't; however, as mature adults we have reached stasis, and our families and doctors would be very concerned if we continued to get taller. McKibben argues, and I agree, that our economies are now mature.

At best science may be a bandage on an unsustainable, growth-based economy, but it won't be the cure. Our economies and societies are complex ecosystems constrained by thermodynamics, and from thermodynamics we know that all this growth requires energy and generates waste (entropy). According to Philip Ball, former *Nature* editor, "creating a true science of sustainability is arguably the most important objective for the coming century; without it, not an awful lot else matters. There is nothing inevitable about our presence in the universe."[8]

Throughout the 20th century, children enjoyed better lives than their parents. However, this pattern cannot continue forever; something has to give: the economic system built on growth, the environment, our energy consumption, and/or our eating habits. There is no denying that an increase in scientific knowledge will increase the quality and length of our lives, but it will also bring increasing environmental and ethical problems. It will widen the gap between the haves and have-nots, both within the United States and between nation-states. Developments in science won't just improve transportation (cars, trains), communication (phones, internet), and consumption (fertilizers), as they did in previous generations; instead new advances have the potential to improve our bodies (CRISPR; chapter 9) and minds (optogenetics; chapter 8).

In *Homo Deus: A Brief History of Tomorrow*, Yuval Harari argues that existing inequities will be compounded by the fact that in the past, medicine's main purpose was to heal the sick, whereas in the future medicine will increasingly be designed to enhance the healthy. Treating the sick is an egalitarian process, while upgrading the healthy will be a luxury available only to the elite. Medicine will increase existing inequities by giving an edge to the rich. "People want superior memories, above-average intelligence and first-class sexual abilities. If some form of upgrade becomes so cheap and common that everyone enjoys it, it will simply be considered the new baseline, which the next generation of treatments will strive to surpass."[9] For the first time, the rich will have not only significant material benefits but also genetic improvements. As in the past, they will have better lives than the poor, but thanks to new techniques such as CRISPR, they may actually *be* better, too. In the extreme case, this could ultimately lead to a new species: *Homo superior*. Today the richest 100 people have more assets than the poorest 4 billion. In the future this financial inequality may lead to biological inequalities.[10]

ETHICAL AND SAFETY CHALLENGES OF SCIENTIFIC GROWTH

Authors and thinkers, including Ray Kurzweil and Bruno Giussani, suggest that science and technology are growing exponentially, while the structures of our society (government, education, economy, etc.) are designed for predictable linear increases, which are dysfunctional in today's exponential growth.[11] This is why our nation-state system can't deal with the challenges of modern science. The challenges presented by modern science (climate change, CRISPR, gene drives, and artificial intelligence) are much larger than those brought about by the Industrial Revolution (steam engines and electricity). Even if we find ways of globally regulating science, there will always be a country marketing itself as a place to do research that is banned elsewhere. And it just takes one country pursuing a high-risk, high-profit path for all the other countries to follow. In fact, the nation-state/growth economy that exists today requires that countries follow such paths to avoid falling behind.

Many governments, including that of the United States, control research by intentionally not funding certain areas that are either dangerous and unethical or difficult to regulate. This technique doesn't work when foundations or start-up companies fund the work. It also fails when the techniques and materials being used are inexpensive, as is the case with CRISPR (chapter 9), and government funding isn't needed (The Amateur Scientists; chapter 3). In the absence of clearer guidelines or regulations, scientists have to rely on themselves, on their own scientific norms. This doesn't work too well in modern science because of the intensely competitive nature of academia, in which "the drivers are about getting grants and publications, and not necessarily about being responsible citizens," notes Filippa Lentzos of Kings College London, who specializes in biological threats.[12] High-profile results matter. In addition, to prevent their competitors from knowing what they are doing and prevent being scooped, scientists keep their experiments under wraps until they are ready to publish, at which point the cat is let out of the bag, and it is too late to think about the ethical impact of the work or to try to stop the research.

Dual-use research, which could be used for either good and ill, presents its own challenges to the safe and ethical regulation of global scientific research. Occasionally scientists work their way to an invisible dual-use research line and cross it. In response, surprised, shocked, and scandalized scientists have urgent meetings to discuss the moral and safety implications. Scientists often proudly point to the 1975 Asilomar conference on recombinant DNA as a model response for dealing with science that has reached new and challenging boundaries in ethics and safety. They perceive the conference as a successful self-regulation in the public's interest. Senator Ted Kennedy and other politicians of the time saw it differently; they considered the scientists a group of unelected experts making public policy without public input. Scientists need broader input from the general public and ethicists, but they are hamstrung by the goals and modus operandi of the expert collaborators they need. Philosophers and ethicists take a contemplative, long-term perspective, while engineers are eager to take results from the laboratory to the market, and investors are always in a hurry and looking for short-term financial gains. Consequently, we have been very good at commercializing scientific discoveries but less proficient at predicting their consequences and proposing the appropriate guidelines (e.g., DDT, fracking, nuclear chemistry). The increasing speed at which scientific breakthroughs are being made will also make it harder and harder to predict and regulate them in the future.

Scientists, despite their desire to have inputs into policy related to their community's discoveries, are not trained to anticipate the consequences of their research, and their solutions are often ineffective, as evidenced by the frequency of such "transgressions' and mini "Asilomars." For example, in 2002 scientists from the State University of New York, Stony Brook synthesized a polio virus from scratch; in 2005 researchers from the Centers for Disease Control and Prevention (CDC) reconstructed a particularly virulent form of the 1918 flu virus; in 2012 two teams mutated the bird flu virus to make it more virulent in mammals; in 2017 a group at the University of Alberta resurrected a horsepox virus, a close cousin of the smallpox virus; and in 2018 CRISPR was used for the first time to create genetically modified human babies. Each of these experiments crossed a line that may have unforeseen consequences, and each led to an emergency conference. Each case leads us closer to the point at which one small accident or well-placed malicious scientist can affect a large portion of the human population or even accidentally wipe out an entire species. In an interview with the *Atlantic*'s Ed Yong, Kevin Esvelt, a CRISPR/gene drive expert at MIT, succinctly summarizes the problem: "Science is built to ascend the tree of knowledge and taste its fruit, and the mentality of most scientists is that knowledge is always good. I just don't believe that that's true. There are some things that we are better off not knowing."[13] On the other hand, we have to remember that some research, such as in vitro fertilization, was once seen as a transgression of scientific norms but is now scientifically and socially acceptable.

CHALLENGES TO SCIENCE

President Donald Trump is not a strong supporter of science, the scientific method, or facts. In his tweets he promotes conspiracy theories and implores Americans to distrust conventional sources of information and traditional institutions. In his first 5,000 tweets as president, the words "science" and "technology" were never used. It took two years before he appointed a science adviser. Deregulation has been a priority of the Trump administration, and by July 2019 it had rolled back more than 80 environmental rules and regulations.[14] It has weakened the Environmental Protection Agency (EPA), cutting staff and budgets and undercutting the agency's ability to use science in its policy making, resulting in steep drops in civil and criminal enforcement of violations of laws such as the Clean Air and Clean Water Acts.[15] According to the EPA, the number of "unhealthy-air days" has increased by 14 percent under this administration, and ozone, nitrous oxide, and particulate matter are more common than in 2016.[16] One of the most telling facts is that both the EPA and CDC have been prohibited from using the phrase "evidence-based" in their publications and press releases. So far, thanks to congressional intervention, the budgets of the National Science Foundation (NSF) and the National Institutes of Health (NIH) have survived, despite three requests by the Trump administration to reduce their size and their budgets.

Ironically, although he has claimed that climate change is "bullshit," "pseudoscience," and "a total hoax," President Trump's representatives have applied for permission to erect a sea wall to protect one of his golf courses in Ireland from rising seas due to "global warming and its effects."[17] (Although this book could easily have been a rant about the Trump administration and its attitudes toward science, I have tried to show some restraint and have limited my comments about the president to the first and last chapters.)

President Trump's opinions weren't formed in a vacuum. He was elected by the American public and still has support. This reflects the increasing public mistrust and resentment of experts. This rejection of scientific thinking and evidence comes from many directions: postmodernist academics and journalists, Christian fundamentalists, liberal new-age purists, and industrial interests and lobbyists. A 2015 Pew Research Center poll showed that "a sizable opinion gap exists between the general public and scientists on a range of science and technology topics," and that "compared with five years ago, both citizens and scientists are less upbeat about the scientific enterprise."[18] In *The Workshop and the World: What Ten Thinkers Can Teach Us about Science and Authority*, Robert Crease writes, "Some people, including many scientists, seem resigned to this. They hope that scientific authority is a natural thing that will shortly reassert itself, like a sturdy self-righting boat knocked over by a rogue wave." He argues that this is not going to happen because the scientific process described earlier in this chapter is inherently vulnerable to attacks. "The fact that it is done by collectives, is abstract and always open to revision" provides fuel for science deniers. To change their minds, we can't just explain the science over and over again; we have to learn how they think and why they are rejecting science.[19]

Many scientists and science supporters have rallied against the antiscience bias, climate denial, flat-earthers, and anti-vaxxers. For example, the first March for Science was held on April 22, 2017; many scientists ran for office in the 2018 elections; and there have been many initiatives to improve scientific outreach.

FAKE NEWS AND SCIENCE

In an essay in the *New York Times Magazine* in 2016, Jonathan Mahler writes that people "are abandoning traditional sources of information, from the government to the institutional media, in favor of a D.I.Y. approach to fact-finding" and are "forming a radical new relationship between citizen and truth."[20] In addition, over the last decade science has been revolutionized by the development of new techniques that allow scientists to conduct experiments bordering on the fantastic, increasing the difficulty for the layperson to distinguish between fact, hyperbole, quackery, and fake news (chapters 9 and 11).

Fake news and pseudoscience occasionally get the better of scientific facts in Congress, too. Congress also struggles with the fact that the amount of scientific knowledge in the world is not only increasing but growing faster and faster. At the same time, science is becoming more complex thanks to a spike in interdisciplinary work between previously disparate fields, such as optics, electrical engineering, and neuroscience joining forces in optogenetics (chapter 8). This has resulted in an ever-widening gap between the scientific knowledge of legislators, religious leaders, and voters and the total available science knowledge.

In Congress, which is ultimately in charge of regulating and defining the direction of science research in the United States, these difficulties are amplified by the fact that there are only 3 scientists and 8 engineers in the 115th Congress of the United States, while there are 218 lawyers. The 7 radio talk show hosts, 26 farmers, and 8 ordained ministers all outnumber the scientists as well. Similarly low numbers are found in Australia and Canada, where scientists make up just 4 percent of each country's parliament. The vast predominance of lawyers in the House of Representatives and Senate sets the tone of the debate in the U.S. Congress. Trial lawyers are trained to win debates, they use facts selectively, and they aren't looking for the truth, nor are they interested in presenting the whole picture. In contrast, science relies on gathering evidence, weighing that evidence, and validating theories.[21] Scientists and science in general don't do well in politics (Angela Merkel and Margaret Thatcher, both chemists, are obvious exceptions). Scientists believe in the importance of facts and think they can win public debates by using facts, despite empirical evidence that suggests passionate opinion will often overcome scientific facts. We can no longer rely on Congress to provide the leadership and guidelines for scientists and industries to deal with the problems and ethical dilemmas associated with gene editing (chapter 9), climate change (chapter 11), and quackery (chapter 12). Scientists have to become more media savvy. They have to learn how to interact with journalists, regulators, and politicians, and they need to have a larger presence on social media. Scientists make good administrators, and many are university presidents; it is time some make the transition into politics.

Having placed today's science in a wider context, it is time to see the *new* science, contrast it with the *old* science, see all that *good* science can do, and lament how it can be abused as *bad* and pseudoscience. (In an earlier incarnation, this book was subtiitled "New Science, Old Science, Good Science, Bad Science.")

Chapter Two

The Professional Scientist

Is the disheveled, gray-haired, Einstein-like character in a lab coat still a good representation of a scientist? Who are our scientists, and who should they be?

In an interview with the *Guardian*, Donna Strickland, 2018 physics laureate and the third woman to ever receive a Nobel Prize in physics, says, "I don't see myself as a woman in science. I see myself as a scientist."[1]

As discussed in this chapter, the situation for women and people of color in science may have improved over the last decade, but inequities still exist. Many people still see the woman before they see the scientist. The United States and the world as a whole have not been taking full advantage of the diverse pool of potential scientists. We may have found and nurtured many future Einsteins, but we have fallen behind in cultivating new Marie Curies and George Washington Carvers. To stay competitive in the current economic system, to solve our global food needs, and to overcome our environmental problems, countries, companies, and academic institutions need to make use of all scientific talents available across a vast array of gender identities, races, and ethnicities.

From the 1970s to 2019, the number of current college graduates has flipped from being 58 percent men to being 56 percent women.[2] However, the gender distribution is not uniform; while women receive 59 percent of bachelor's degrees awarded in the biological sciences, they receive only 40 percent of physical science and mathematics degrees and much less than 20 percent in the computer sciences and engineering.[3] Women make up half of the total U.S. college-educated workforce but only 29 percent of the science and engineering workforce. A 2017 National Center for Science and Engineering Statistics report shows that although white men make up only one-third of the U.S. population, they constitute at least half of all scientists.[4]

SCIENTISTS OF COLOR

Although students from underrepresented groups aspire to careers in science, technology, engineering, and mathematics (STEM) fields at the same rates as their nonminority peers, minorities, who comprise 30 percent of the U.S. population, make up only 14 percent of master's students and just 6 percent of all PhD candidates.[5] This gap hasn't changed much in the last 15 years.[6] In 2017, there were more than a dozen areas in which not a single PhD was awarded to a black person, primarily within the STEM fields.[7] There are many reasons for this. In a paper examining underrepresented minority participation in biomedical research and health fields, Rosalina James, a member of the University of Washington Bioethics and Humanities department, states, "Inadequate preparation is a major limiting factor in efforts to increase the pool of qualified minority applicants for advanced education. Poverty, sub-par resources in minority-serving schools and poor mentorship contribute to losses of minority students at each level of education."[8] Stereotype threat,[9] impostor syndrome, and numerous microaggressions[10] also prevent scientists from minority groups from performing anywhere close to their potential.

According to a report by the American Institute for Research, a third of all black STEM PhDs earned their undergraduate degrees at historically black colleges and universities (HBCUs), institutions of higher education founded to serve primarily African American students.[11] Xavier University of Louisiana, located in New Orleans, is an HBCU and Catholic institution nationally recognized for its STEM programs. "Of the 3,231 students enrolled at Xavier in fall 2018, approximately 72 percent were African American, and about 79 percent of the 2,463 undergraduates majored in the biomedical sciences (bioinformatics, biochemistry, biology, chemistry, computer science, data science, mathematics, neuroscience, physics, psychology, public health sciences, and sociology)."[12] Xavier is best known for its education in the health professions, and, according to 2018 *Diverse Issues in Higher Education* data, ranks second in the nation in the number of African Americans who earn bachelor's degrees in the physical sciences and fourth in the number earning bachelor's degrees in the biological and biomedical sciences.[13] A 2013 National Science Foundation report confirms Xavier's success in educating science graduates, ranking Xavier first in the nation in producing African American graduates who go on to receive life sciences PhD degrees, fifth in producing African American graduates who go on to receive science and engineering PhD degrees, and seventh in producing African American graduates who go on to receive physical sciences PhD degrees.[14] The 2012 Report on Diversity in Medical Education published by the Association of American Medical Colleges (AAMC) ranks Xavier first in the number of African American alumni who successfully complete their medical degrees.[15] Xavier is one of 101 HBCUs in the United States. I went to visit Xavier to get the university's perspective on minority representation in STEM fields and to see what it does so well.

In 2014, the National Institutes of Health (NIH) announced 10 BUILD (Building Infrastructure Leading to Diversity) awards, ranging from $17 million to $24 million over five years. In his announcement of the awards, NIH director Francis Collins explained that the program was designed to increase the representation of African American, Hispanics, and Native Americans in science. Collins is particularly concerned because "although 12.6% of the U.S. population is African-American, only 1.1% of our NIH principal investigators are African-American."[16] One of these awards, in the amount of $19.6 million, was presented to Xavier.

Professor Maryam Foroozesh is the chemistry department chair and the lead principal investigator of Xavier's NIH BUILD award. Like Professor James, she feels the root of the problem is in K–12 education. The public school system needs serious improvement across the country. If we are all paying taxes, then every one of our children should have the right to the same type of education. The underrepresentation of students of color in the sciences is not due to their ability but to a lack of preparation and the reduced expectations that can come with an inferior education. "I think if there was a standard K through 12 national education program with federal oversight, then all the students including the ones from the inner cities or rural areas of the country would get a better education," Foroozesh told me. "A federal education system would hopefully also address some of the diversity issues you see in science at the higher levels, because once you provide all the students in the U.S. with the education they deserve, then you would get a higher number of scientists coming out of the groups currently underrepresented in science."

Foroozesh is also very concerned that the 116th Congress is extremely focused on short-term returns on their investments. They are pressuring funders to do short-term assessments, but it takes a long time to see the results of diversity programs. Increasing the diversity of the faculty and grant writers is a long-term project that involves K–12 reforms, changes in both undergraduate and graduate programs, and faculty hiring and retention. These factors are interlinked, and changes are difficult to evaluate because the numbers we are talking about are small. In 2012, 267 African Americans and 329 Hispanics received PhDs in the biological sciences. Even a small increase or drop can represent a large percentage change. Foroozesh worries that funding to programs that foster diversity and help STEM undergraduates from underrepresented groups are being cut before the programs have adequate time to prove their worth, which will disproportionally affect HBCUs that are highly reliant on them because they don't have hundreds of millions of dollars in endowments.

There are many reasons students do not continue in STEM, and any one is sufficient to dissuade a student from persevering. Initial assessments of Xavier's BUILD program have shown that mentorship is a major factor. Mentoring is crucial in retention of underrepresented students in STEM fields. Ideally the mentors need to serve as role models and have to understand the importance of cultural issues, family ties, financial needs, and expectations. Xavier's researchers undergo special mentoring training to achieve this understanding.

Seventy-five percent of the undergraduates enrolled at Xavier in fall 2018 were African American women, most of whom were STEM majors. These women in STEM largely fell between two departments at the university. The chemistry department, the largest department at Xavier in the number of faculty and research staff and the second largest in the number of majors, has 28 faculty members, half of whom are women. Even though Xavier is an HBCU, the department only has six African American/black faculty (1 woman and 5 men). The biology department, the second largest department at Xavier based on the number of faculty members and the largest based on the number of majors, has 23 faculty members, 6 of whom are African American/black (4 women and 2 men). The causes for this disparity are systemic and have roots in retention and recruitment, as well as the small numbers of African Americans/blacks, especially women, looking for jobs in academia. Another chemistry faculty member, Professor Mehnaaz Ali, is a coprincipal investigator on a National Science Foundation (NSF) ADVANCE grant, "which is focused on creating an equitable, inclusive and energizing climate for female STEM faculty members by addressing systemic barriers which currently lead to higher attrition rates of female faculty and women of color." Dr. Ali told me, "There are many speed bumps for female faculty in academia. It could be child care, it could be family care. If you are a minority in a school such as this one, where the numbers are low, you end up doing a lot of service, if you are a female faculty of color you are a role model for everyone, you are on a lot of service-oriented committees. But are you on the committees that are high impact? This is significant because committees, mentoring, and other unseen burdens result in burnout that can lead to retention issues."

Another reason Xavier has such high graduation rates for African American women in science is that it has a critical mass of students of color doing science. The students have each other, they can talk to each other, and they look to each other and Xavier alums as role models in science. This can make all the difference in the world. The critical mass required for this type of group-wide support is absent in most non-HBCUs. The ADVANCE team aims to create similar environments for faculty to have a safe space where they can talk and address important issues with people who are like minded and share intersectional boundaries and thus be connected to a larger campus-wide faculty network.

I grew up in South Africa. While at university I tutored students from Soweto who were boycotting their apartheid education. Each Saturday they traveled more than two hours to learn math and science from a naive undergraduate, who was younger than most of them. This experience made me realize how important education is and the lengths some people go to get it. At Connecticut College I do a lot of chemistry outreach and get to see the different levels of preparation students receive. Based on 25 years of personal experience, I argue that the inequities (facilities, class size, and equipment) between the richer and poorer schools that I visit in the Northeast are growing worse. In 2007 I started, and today still direct, an undergraduate program that prepares students from underrepresented groups for a variety of science-related careers and provides a solid foundation for graduate study or medical school. Participants in the Science Leader program come from socioeconomically disadvantaged backgrounds, and priority is given to students of color, first-generation college students, students with disabilities, and women in mathematics, computer science, and physics. The program is based on cohort formation through a first-year seminar, an associated field trip, and research. As students go through the college acclimation experience together, they quickly become part of the larger Science Leader community. They learn from advanced students about what to expect in certain courses. Older students organize study groups, peer tutoring sessions, and social gatherings and assist with the orientation of incoming students. This creates a supportive network of science professionals and graduate and undergraduate students that grows and strengthens organically. Since 2007, 104 students have entered Connecticut College under the Science Leader program, an average of approximately 15 students per year. The six-year graduation rate for students in the Science Leader program is 97 percent. As of spring 2019, Science Leader students have obtained six medical degrees, one doctorate, and eight master's degrees in STEM fields and six other graduate degrees. Twenty-four Science Leader alums are currently enrolled in graduate schools. The Science Leader program has been named a recipient of *INSIGHT into Diversity* magazine's 2019 Inspiring Programs in STEM Award.

I wish we didn't need this program, but it has become increasingly important and relevant. Most universities have been forced to introduce similar initiatives. The program is both one of the most rewarding and most depressing facets of my job.

NOBEL PRIZES

What Do They Say about Diversity?

As I write this we have just gone through another Nobel Prize season. The three Nobel Prizes in chemistry, physics, and medicine are the scientific equivalents of the Academy Awards. Just as for the Oscars, there is a pre-Nobel buzz; scientists are trying to predict who will be awarded the year's prizes. In the days and weeks following the announcement of the awards, there is a thorough analysis of the winners and their research and sympathizing with those who, unjustly of course, didn't get an award. It doesn't take a very detailed investigation to discover that women and black scientists are not proportionally represented among the laureates, the United States does better than most countries, and China has surprisingly few science Nobel laureates.

In 1895, five Nobel Prizes were established according to Alfred Nobel's will. The first prizes in chemistry, literature, physics, and medicine were awarded in 1901. Each prize can be awarded to no more than three people, and prizes may not be awarded posthumously.

The annual Nobel announcements occur in October, which is Black History Month in the United Kingdom. This is rather ironic, as no black scientist has ever won a Nobel Prize in science. Zero of the 617 STEM laureates! The reasons for this are the limited opportunities black (especially African) students have and the biases, hurdles, and lack of role models that they experience in science. Not enough young black students are choosing science, and there are not enough black full professors in the sciences at elite universities, where the networks and reputations required for winning a Nobel are made. Unfortunately, it is impossible to give Africa the same economic and political power as the global North. But "if we want more black scientists and eventually Nobel laureates, then similar direct strategic action (as has been used to increase the numbers of women in science) is urgently needed."[17]

Immigrants to the United States

Thirty seven of the eighty-nine U.S. citizens awarded a Nobel Prize since 2000 were foreign born. Most notably, all six American winners of the 2016 Nobel Prize in economics and STEM fields were immigrants to the United States.

American universities consistently perform extremely well in all global rankings of academic institutions. Foreign graduate students flock to the United States. In 2015, more than half the computer science, engineering, mathematics, and statistics graduate students were international students. Most of these students return to the countries of their birth upon completing their graduate studies, but a significant number of the very best stay in the United States and become naturalized citizens. In a disturbing trend, the NSF reports that the number of international graduate students coming to the United States dropped by 22,000 (5.5 percent) in 2017.[18] *Inside Higher Education* reports that the high cost of US higher education, visa denials and delays, the political and social environment in the United States, and increasing competition from other countries are responsible for this decrease.[19] Had the proposed tax on graduate fellowships passed Congress (it barely failed in 2017), the decrease would surely have been greater than just 5.5 percent. In 2018, the number of international students dropped again, especially in universities in the central parts of the United States and at lesser ranked universities. The numbers of students coming from Canada and Mexico also declined.[20]

What will this drop in international STEM graduate students, often the best from their countries, mean to science research? This change in demographics does not bode well for science in the United States. We need to be careful we don't lose touch with this very important talent pool. Not only do immigrants contribute to an inordinately high number of Nobel awards, but they also bring new ways of thinking to their research labs. They come from other cultures and have learned their science in different educational systems, which place different emphases on rote learning, historic understanding, and interdisciplinary research. They often bring an alternative and important perspective that a homogeneous scientific community cannot match.

WOMEN IN SCIENCE

Between 1901 and 2019, there were 213 Nobel awardees in physics, 184 in chemistry, and 219 in medicine. Over that period women were only awarded 3 Nobel Prizes in physics, 5 in chemistry, and 12 in medicine.

Donna Strickland was awarded the 2018 Nobel physics prize with her PhD mentor, Gérard Mourou. Strickland was the first woman to be awarded a physics prize in 55 years. At the University of Rochester, Strickland and Mourou together developed the most intense and shortest laser pulses ever produced in a laboratory. Mourou had the idea, and Strickland made it work. Besides being an impressive scientific advance, their technique has resulted in high-intensity lasers that have been used in millions of corrective eye surgeries.[21] Ironically, Donna Strickland wears glasses and refuses to get the laser eye surgery that her research made possible: "I have great faith in lasers, but no one's putting one near my eye."[22]

Many women could have, and probably should have, been awarded a Nobel Prize in physics. The *Guardian* published a 2018 article titled "Five Women Who Missed Out on the Nobel Prize."[23] Lise Meitner, who laid the groundwork for understanding nuclear fission, is my favorite of these. An element, meitnerium, was named for her posthumously. She is the only woman to have earned such an honor (curium is named after both Marie and Pierre Curie). But no Nobel. Both Lise Meitner and Jocelyn Bell, who discovered the first radio pulsars in 1967, missed the Nobels while their male collaborators, Otto Hahn (1944) and Anthony Hewish (1974), were each honored with an award. Lene Hau, a physicist at Harvard University, is another woman mentioned in the *Guardian* article. In 1999, her team was able to slow a beam of light to 17 meters per second, which she topped in 2001 by stopping a beam of light completely. This work has implications for quantum computing and quantum encryption. Hau's work is fairly recent, so she may yet get a Nobel Prize.

There can be no denying that just three women in 213 physics Nobel laureates is a disproportionately low number and that many distinguished and immensely qualified female physicists must have been overlooked. But is this a big deal? Yes, of course it is. It is grossly unfair to the women who didn't get the award and sends the wrong message to young people, funding agencies, editorial boards, and others about who does noteworthy science. Perhaps much more important, it is indicative of many biases and inequities that plague women and minorities in science.

In 2008, I served as a consultant for the Royal Swedish Academy of Sciences' deliberations about the chemistry award; as a result, my wife and I were invited to attend the Nobel ceremonies. We stayed in the Grand Hotel with all the awardees. We got to see how scientists, excellent but unknown outside their fields, suddenly became superstars. They were interviewed on radio and television and hobnobbed with Swedish royalty. The events of Nobel week were shown live on Swedish television, and the newspapers were atwitter about the clothes worn by the Swedish princesses at the awards ceremony. Nobel laureates immediately become role models who are invited to give seminars all around the world. In an interview with *Nature* magazine, Donna Strickland, was asked how her life had changed since being informed that she had won the award. She said, "Oh, completely! This is just completely crazy, you know; I got to talk to the Prime Minister of Canada for the first time ever. He was very nice about it. I said, 'This must be how your life is like all the time.' And he replied, 'No, I don't always get to speak to a Nobel laureate.'"[24] Her answer shows the stature the prize imparts and why women Nobel laureates are such important role models.

Because only 3 percent of the science awardees have been women and there have been no black winners, there are very few role models for the new generation. The entertainment industry is no help; media depictions of male scientists and engineers outnumber those of women by a ratio of 14 to 1.[25] Frances Arnold, winner of the 2018 Nobel Prize for chemistry, had a guest appearance on *The Big Bang Theory*. She was the first woman scientist to make a guest appearance in 12 seasons of the show. If we want to solve our climate change problems, cure Alzheimer's, expand our economies, and so forth, we can't afford to completely ignore a large proportion of the population. In the words of Virginia Valian, who has spent the past 25 years studying the structural and psychological reasons for the paucity of women in the upper reaches of science, "If we want talent, we have to welcome it and nurture it, in all its diversity."[26]

Nomination to receive a Nobel Prize in science or medicine is by invitation only. Each year, thousands of members of academies, university professors, scientists, previous Nobel laureates, members of parliamentary assemblies, and others are asked to submit candidates for the Nobel Prizes for the coming year. The names of the nominees and other information about the nominations cannot be revealed until 50 years later.[27] Despite this confidentiality, we know that nominations tend to favor scientists working at elite research institutions, famous scientists who are good at self-promotion and are well known to their peers. Predictably, these tend to be older, established white males. The Royal Swedish Academy of Sciences, for chemistry and physics, and the Nobel Assembly at the Karolinska Institute, for medicine, are in charge of selecting the Nobel winners from the nominations. They are aware that they have a "white male problem," and starting with the 2019 nominations they have asked nominators to consider diversity in gender, geography, and topic in their future nominations. It didn't work. There were no female awardees in physics, chemistry, or medicine at the December 2019 Nobel award ceremonies.

"The Leaky Pipeline"

The disproportionately low number of female Nobel laureates in the sciences and the absence of black science laureates is an extreme example of the "leaky pipeline" in science. The NSF coined the phrase for this phenomenon in the 1980s. It comes from a report in which the NSF also predicted an upcoming shortage of scientists and engineers that would grow to over 500,000 by 2006. The shortage never materialized, but the metaphor stuck. It presents a vivid visual image of women and people of color entering the sciences but then "leaking out of the pipeline" at greater rates than white males as they progress along their educational and career paths. This pipeline should lead to awards and board memberships in science, the ultimate being the Nobel Prizes, but the number of women and people of color consistently decreases as we move along the pipeline. More than 14,000 academic articles have been written about the "leaky pipeline."[28] However, while the phrase is also very popular with the media and politicians, it is a flawed metaphor. Although it is often used in connection with a perceived future shortage of scientists, it is less commonly used to show the need for increased diversity and equitable representation in STEM, the more important pressing problem. At the same time, a leaky, dripping pipeline has the obvious negative connotations of a dysfunctional pipe; it implies that there is a single direct pathway from preschool to PhD, and that the PhD is a more valuable endpoint than other educational degrees.[29] Although the leaky pipeline is not necessarily a good metaphor, it certainly describes something real. Therefore, I use the phrase in this chapter, although I place it in quotes to acknowledge its inadequacy.

The percentages of women decrease from a bachelor's degree to a PhD, tenure, full professorship, and major awards in the sciences. A 2015 NSF report shows that women accounted for 45 percent of PhDs in the STEM fields. The percentage falls to 42 percent for female junior faculty members and to 30 percent for female senior faculty members.[30] A similar drop-off is observed in biotech companies, where women just hold 20 percent of leadership roles and 10 percent of board seats. *Chemistry & Engineering News* reviewed the boards of 75 biotech companies that had raised series A funding since 2016 and found that 39 have only men on their boards, while just 2 have boards with over 30 percent women.[31]

There are numerous reasons for the decrease in the percentage of women in more senior positions and receiving major awards. The remainder of this chapter divides them into three main categories: (1) a structure (PhD, tenure, funding, and publication) that is not compatible with a family life; (2) implicit biases against women by other scientists (both men and women); and (3) a system that favors men and masculine confidence.

Science and a Family Life

The easy and convenient explanation for the low numbers of women in the upper levels of science is that they have more family responsibilities . Many will even argue that this is a fait accompli and that nothing can be done about it. I disagree. Some of the differences may be due to family reasons, but with proper incentives these differences can be made negligible, and there are other, more significant factors that cause women to exit the pipeline. If family issues are the only problem, why have the last 30–40 years seen such great improvements in gender diversity (and even racial diversity) in the life sciences, while the physical sciences, computer science, and engineering have lagged behind? Physics and astronomy require very similar skills, yet astronomy has twice the percentage of women faculty as physics.[32]

A study of gender diversity in the life sciences sector in Massachusetts was conducted by Liftstream and MassBio, in which over 900 people working in the biotech sector were surveyed. The 2017 report found that women have career breaks more often than men, related not just to parenthood but also to caring for elderly parents. More important, the researchers found that parenthood isn't the only cause for the "leaky pipeline." In fact, more women leave the biotech sector because they are opting out of the corporate culture than for parenting reasons.[33]

Patricia Fara is the president of the British Society for the History of Science and a fellow at Clare College, Cambridge. She has an undergraduate degree in physics from Oxford but is one of the many who leaked out of the "pipeline." In a February 2018 National Public Radio interview, she talked about why she had dropped out of the system. For her it was a choice between quality of life and status. To succeed in science and academia routinely requires a 24/7 commitment. Fara feels that she and many other women have wisely opted for a better quality of life and that "perhaps in time, the really smart men will realize that's a better option than earning more money but having no time to spend it."[34] She might be right that faculty at the elite institutions have little or no life outside of work and that getting tenure requires extraordinary sacrifices, especially in one's family life. Progressive policies such as paid parental leave, high-quality, on-site child care, and tenure 'clock stops" will improve the quality of living of STEM faculty and make STEM careers more appealing to new generations, but they won't completely close the gender gap.

In their aptly titled paper "Do Babies Matter? The Effect of Family Formation on the Lifelong Careers of Academic Men and Women," Mason and Goulden have shown that women with children don't advance any slower than women without children. That doesn't mean having babies doesn't matter; it matters a great deal. The study showed that there is large gap in achieving tenure between women and men who have babies within five years of getting their PhDs. But most important, based on all the data in their study, Mason and Goulden conclude that "babies are not completely responsible for the gender gap, and that there are other factors at work, perhaps including the thousand paper cuts of discrimination."[35] Most of these "paper cuts" are a result of implicit bias. They are unconscious, involuntary, natural, and unavoidable assumptions that all of us make on the basis of subconscious assumptions, preferences, and stereotypes.

Implicit Bias

Gender disparities are decreasing in academia. However, many biases and gender inequities remain. This section looks at some of these "paper cuts." Frances Trix and Carolyn Psenka of Wayne State University examined 300 letters of recommendation for medical faculty. They found significant differences between letters written for men and women. The average length of letters for female applicants was 227 words, whereas the average length of letters for male applicants was 253 words. Not only are the letters for women shorter, they also use descriptors such as "determined" and "dependable" more often and "outstanding" and "brilliant" less often than the equivalent letters for men. Letters for women are more likely to mention family situations and personal characteristics. And here is the kicker: it makes no difference whether the letters are written by men or women.[36]

The peer-review publishing model that the scientific publication system is based on is a single-blind process. Upon receiving a manuscript, a journal editor sends it to external reviewers with expertise in the research area. The reviewers know the identity of the authors, but the reviewers remain anonymous. They read the paper, recommend whether it should be published or not, and identify what changes are needed to make the paper acceptable if it is not quite ready for publication. To examine reviewer bias, Silvia Knobloch-Westerwick, Carroll Glynn, and Michael Huge of Ohio State University[37] recruited graduate students to rate conference abstracts authored by researchers with distinctively male or female names. The fake author identities on the abstracts were varied such that the same abstract would be attributed to a male-sounding name or a female-sounding name in a given test. Scientific abstracts submitted by "male" authors were considered of higher scientific quality than those submitted by authors with feminine names even though there was no difference in content. The gender of the reviewers did not influence these patterns. The differences were small but statistically significant. I am confident the same implicit biases appear when papers or presentations indicate that a work was done at a lesser ranked institution or in a developing country, or if the author has a foreign last name. Bias in peer review can affect the publication record of young scientists and impact their chances for promotion and tenure, a painful paper cut indeed.

The reviewing biases discussed here are not limited to graduate students. In 2012, Jo Handelsman and coworkers at Yale University showed that faculty at research-intensive universities favor male students. In a randomized double-blind study, 127 science faculty rated the application materials of a student, randomly assigned either a male or a female name, for a laboratory manager position. They found that "faculty participants rated the male applicant as significantly more competent and hireable than the (identical) female applicant. These participants also selected a higher starting salary and offered more career mentoring to the male applicant. The gender of the faculty participants did not affect responses, such that female and male faculty were equally likely to exhibit bias against the female student."[38]

In 1998, Virginia Valian published *Why So Slow?*, a landmark book on bias.[39] In 2018, she and her coworkers analyzed gender differences in 3,652 colloquium speakers who presented their work at 50 prestigious U.S. colleges and universities in 2013–2014.[40] The proportion of women presenting colloquia was significantly smaller than for those presented by men. There was no difference in the extent to which male and female professors at these elite universities valued or declined speaking invitations. The difference was in the number of invitations offered. These biases have significant consequences, because as the authors say, "Colloquium talks are an important part of academicians' careers, providing an opportunity to publicize one's research, begin and maintain synergistic and productive collaborations, and enhance one's national reputation; those results in turn typically lead to retention, promotion, or greater salary increases. . . . Colloquium talks also signal to audience members who is worthy of being invited."[41]

Ending implicit biases is not going to be easy. Combating implicit bias is difficult at the best of times, but it is particularly hard in the sciences, where scientists believe that the process of doing science is rigorous and objective and as a consequence are convinced that they are not prone to bias. "Gender discrimination is everywhere," says Christine Williams, a sociologist at the University of Texas at Austin. "But what makes the experience unique among scientists is their almost unflappable belief in objectivity and meritocracy."[42] Another complication is the fact that in acknowledging implicit bias against underrepresented groups, established white male researchers have to accept that they may have been privileged in the attainment of their positions. "Some scientists might be slow to consider that the system could be rigged because it implies that their own accomplishments might not be totally deserved," says Deborah Rhode, a legal ethicist at Stanford University. "They might also be less willing to see how helping their closest peers (mainly males) might simultaneously exclude others."[43]

Biases can lead to discrimination, a much deeper cut than implicit bias. A 2018 Pew Research Center report finds that the majority of black people in STEM fields (62 percent) report having experienced some form of discrimination at their work due to their race or ethnicity. The survey also finds that half of women working in STEM jobs report experiencing discrimination at work due to their gender, more than women in non-STEM jobs (41 percent) and far more than men working in STEM jobs (19 percent).[44] As mentioned previously, more women working in the Massachusetts biotech sector left their places of employment because of workplace issues than for family reasons. Discrimination does not lead to an inviting workspace, and it encourages scientists with important ideas and skills to leave the field.

SCIENCE AND MASCULINE (OVER)CONFIDENCE

Society, and science specifically, rewards masculine (white American) confidence. Numerous studies have shown that in mixed-gender groups, men talk more than women, and that when women do speak they are more likely to be interrupted than men. In contrast, women are considered rude and abrasive if they interject when men are speaking. These behaviors all add to the perceived influence of men.[45] Science is no different.

More than 20 peer-reviewed studies have shown that men self-estimate their own intelligence higher than women do. Most recently, Katelyn Cooper, Anne Krieg, and Sara Brownell of Arizona State University collected data from their upper-level physiology courses. Sixty-six percent of all males in the class thought they were more intelligent than the average classmate, while 54 percent of women thought they were above average. Of course, statistically 50 percent should be above and 50 percent should be below the average. Not surprisingly, Americans tend to have a higher self-estimation than other nationalities. This was also reflected in the study, with only 46 percent of nonnative English speakers considering themselves above average.[46] When working with partners or in groups, the women were asked to rate themselves relative to their closest classmates. They rated themselves smarter than 33 percent of their collaborators, while men thought they were more intelligent than 66 percent of the group. The study shows that the men in the physiology class overestimate their credentials and skills, while underestimating the intelligence of their female classmates, noting that, as a result, "women themselves doubt their abilities—even when hard evidence such as grades say otherwise."[47] It is worth noting that these observations were made in a physiology/biology classroom setting, because, as Brownell says, "Unlike the more male-dominated fields like engineering and physics, biology is seen as a safe place for women."[48]

Science and *Nature* are the most prestigious scientific journals. Publishing in these journals is a sign that you have made it to the top of your field. The author listed first is typically the PhD student who has done the majority of lab work, while the author whose name appears last in the listing of authors associated with the paper is the one who directed the research: the head of the lab. Ione Fine is a neuroscientist at the University of Washington. When she and her coworkers analyzed the gender of the first and last authors of 166,000 articles published in high-profile journals between 2005 and 2017, they found a new disturbing "leaky pipeline." The proportion of women drops dramatically from PhD students in neuroscience (55 percent), to tenure-track faculty (29 percent), to first author in *Nature* or *Science* (25 percent), to full professor (24 percent), and finally to last author in *Nature* or *Science* (15 percent). In a related large-scale analysis based on over eight million papers across the natural sciences, social sciences, and humanities, it was found that women are also significantly underrepresented as authors of single-authored papers.[49] In an article titled "Perish Not Publish? New Study Quantifies the Lack of Female Authors in Scientific Journals,"[50] Fine and Shen lay the blame for these findings on proven biases in the publishing pipeline. They cite studies showing that in situations in which women have done all the lab work, they are less likely to be given the prestigious first author position;[51] the honor is more likely to go to more assertive male students. In the article, Fine and Shen also point out that these biases become particularly important in the most selective journals that accept only the most superlative and brilliant papers, adjectives more likely attributed to white men than to women or men of color.[52] All these biases make women, who tend to be more restrained, less likely to submit articles to the highly selective journals, particularly as rejection by a journal and subsequent resubmission to a less selective journal can result in one's research being scooped by other groups. In conclusion, Fine and Shen suggest that "the scientific community should demand that journals collect data about gender and ethnicity for article submissions and acceptances, and these data should be publicly available." And more important, they should adopt mandatory double-blind reviews.[53] This solution has helped in the famously gendered world of elite orchestras. Female musicians in the top five symphony orchestras in the United States were less than 5 percent of all players in 1970 but are 25 percent today. The proportion of women hired increased dramatically after auditions were anonymized in the 1970s and 1980s by placing performers behind a curtain. Nevertheless, orchestras are still extremely gendered, not only in general makeup but also in instrument breakdown (similar to the observable difference in the gender breakdown of different STEM fields). In the world's top 20 orchestras, 94 percent of harpists are female, and 100 percent of trombonists are male.[54]

Not only are women invited to give seminars less frequently than their male counterparts, but they behave differently at presentations, asking fewer questions. Alecia Carter, a behavioral ecologist at the University of Montpellier in France, collected observational data at 247 departmental seminars, hosted at 35 institutions in 10 countries. She and her colleagues found that men are 2.5 times more likely to ask questions at colloquia than their female counterparts. Upon further questioning, they established that "women were significantly more likely than men to say that they had kept silent because they were unsure whether their question was appropriate, or because they did not have enough 'nerve' to ask it."[55]

Jennifer Harnden-Koehler, a former executive coach and talent management expert at Talent Acceleration Group, set out to establish why women in midlevel positions at biotech companies aren't promoted to senior-level positions even when they have the required credentials and skills. She found they were having a tough time being seen as strategic thinkers. Women are more likely to present multiple scenarios with backup plans, while men often only offer a single strategy. Women tend to see the backup plans as an important part of leadership, but management interprets the caveats as an indication of indecision. Women tend to say "our plan," while men take ownership, calling it "my plan."[56] Research has shown that this democratic leadership style, typically adopted by women, is long lasting and bridge building.[57] However, Harnden-Koehler has found that biotech companies are more interested in "strong" leaders, leaders with something she calls "loudership."

In May 2018, five months before Donna Strickland was awarded the Nobel Prize, a page about her was submitted to Wikipedia but was rejected because she and her research had not garnered enough internet coverage,[58] another clear example of women not self-promoting as much as men.

CONCLUSION

Eric Lander, founding director of the Broad Institute of MIT and Harvard, wrote an opinion piece in the *Boston Globe* in which he said, "The United States has only 5 percent of the world's population. To stay ahead, we'll need to use all our assets. That means leveling the barriers for women in science and engineering, and closing the participation gap for underrepresented minorities. It also means expanding tech-driven prosperity beyond the two coasts."[59] This is particularly important because the economic center of gravity of the world is shifting as the populations and personal income of Africa and (especially) Asia are increasing. If the United States wants to stay competitive in the world economy, it will have to rely on technological and scientific advances. Science and technology are related to each other, and both will advance faster and further with an expanded and diversified talent pool.

For better or worse, the world economic system is based on growth. In the current system, countries and companies need to expand in order to thrive. Staying the same may be sustainable, but it is not economically desirable. The U.S. agricultural and manufacturing sectors have reached their maximum capacity; they can no longer expand. America's economic growth is predicated on the production/design of new products (iPhones, solar panels, cars, etc.). We need new and improved products, high-tech merchandise enhanced beyond previous models. In other words, the expansion of the U.S. economic system is reliant on scientific knowledge and know-how.[60] The use and insights gained from scientific breakthroughs such as CRISPR, optogenetics, and gravitational waves will keep the United States competitive in tomorrow's economy. To do this we need to maximize our scientific talent.

All the hurdles and biases described here don't apply only to women and scientists of color; they also apply to some white male scientists. Chapter 5 discusses Doug Prasher, who wasn't confident enough to go for tenure at Woods Hole; didn't continue working on green fluorescent proteins because he wasn't being supported; and finally dropped out of science, missing the hundredth Nobel Prize in chemistry in 2008 by the tiniest of margins. His is another, very different, example of a "leaky pipeline": the importance of "loudership" and old-boy network connections.

Frances Arnold, 2018 chemistry laureate, is optimistic. She thinks we may have turned the corner, "as long as we encourage everyone—it doesn't matter the color, gender; everyone who wants to do science, we encourage them to do it—we are going to see Nobel Prizes coming from all these different groups. Women will be very successful."[61] Women perhaps, but I am not convinced that people of color will be fairly represented among STEM Nobel laureates in the next 10–20 years. Unfortunately the systemic, societal, economic, and educational (K–12) differences are too large and too entrenched to expect parity in the next two decades.

Chapter Three

Do-It-Yourself Science

*Citizen Science, the Amateur Scientist,
Biohacking, and SciArt*

Chapter 2 looked at the demographics of today's scientists and the need to increase the proportion of people of color and women in science. These scientists were employed in institutions such as industry, academia, national labs, and research hospitals. The vast majority had graduate degrees in science. A group of very interesting, very important scientists is making a resurgence: amateur scientists. They often have no postgraduate degrees in science and are not employed in the scientific sector. They do their scientific research purely for the love of science. They are amateur in the true sense of the word, the etymology of which harkens back to the Latin *amatore*, which means "lover or friend."

HISTORY OF AMATEUR SCIENTISTS

Amateur science has a storied past. Much of our early science was done by amateur scientists. Here I briefly introduce four of these pioneering amateurs—Michael Faraday, Charles Darwin, Henrietta Swan Leavitt, and Robert Evans—before turning to the new breed of amateur scientists we have come to know in modern science.

Michael Faraday, born in 1791 in Newington, England, had no formal education. He grew up poor and learned to read and write in Sunday school. At age 14 he was apprenticed to a bookbinder. Through reading in his spare time he taught himself about electricity and chemistry. When one of chemist Humphrey Davy's assistants was dismissed for brawling in 1812, Faraday managed to get a position working for Davy at the Royal Institution of Great Britain. Though an amateur in the sense that he had no formal education, Faraday would become one of the greatest experimentalists of the 19th century. Among his many breakthroughs was the invention of the first electric motor and dynamo. He pioneered the field of electrochemistry and discovered diamagnetism and benzene. Some have gone so far as to suggest that Davy's greatest contribution to science was his discovery of Faraday, even though Davy himself discovered five new elements.

Mary Ellen Hannibal, author of *Citizen Scientist: Searching for Heroes and Hope in an Age of Extinction*, describes Charles Darwin as the archetypical amateur scientist: "He did not have an advanced degree, and he worked for no one. He worked for himself—no institution."[1] To be fair, he did have a rich father to support him. The days when amateur scientists such as Darwin and Faraday were revolutionizing science are probably over. However, one area in which amateurs have made and still are making substantial breakthroughs is astronomy.

Henrietta Swan Leavitt was an astronomer in the early 1900s. She was a Harvard "computer," one of many women hired to examine the relative brightness of stars in thousands of photographic plates. The hours were long and the work tedious, and they were paid a pittance. Between 1907 and 1921, Leavitt discovered 2,400 variable stars. She didn't just do tedious, repetitive work; she also discovered a relationship between the period of a star's brightness cycle and its absolute magnitude that made it possible to calculate its distance from Earth.

The record for visual discoveries of supernovae is held by another amateur astronomer. Robert Evans was born in 1937; he is a minister of the Uniting Church by profession. He graduated from the University of Sydney, majoring in philosophy and modern history. Two of my favorite authors have discussed Evans's talents. In *An Anthropologist on Mars*, Oliver Sacks describes Evans's ability to find subtle changes in the starfield as "savantlike."[2] In *A Short History of Nearly Everything*, Bill Bryson presents a great analogy for Evans's ability to detect changes: it is like being able to spot an added grain of salt on a tabletop of salt.[3]

In 1992 Daniel Koshland Jr., the editor of *Science* magazine, published an editorial in which he argued that modern science "can no longer be done by gifted amateurs with a magnifying glass, copper wires, and jars filled with alcohol."[4] To do modern science 1990s style, one needed high-tech equipment, significant funding, graduate and postdoc students, a lab, and a graduate degree in science. It seemed as if the days of the amateur scientist were over. However, Koshland was wrong. Amateur astronomers are still contributing important findings. In the 2010s, amateur astronomers have found 42 new planets, spotted a new dwarf galaxy, and sighted "yellow spaceballs that are a rare view of massive star formation[,] and in 2016 two amateur astronomers filmed an asteroid impacting Jupiter."[5]

THE INTERNET OPENS SCIENCE TO AMATEURS: CITIZEN SCIENTISTS

The growth of the internet, the need for data, low-cost sensor technology, the ubiquitous distribution of cell phones, home computers, and user-friendly mail-order gene editing kits have led to a burgeoning do-it-yourself (DIY) science movement. Amateur scientists have spawned biohackers and citizen scientists.

In 2014, the *Oxford English Dictionary* added the term "citizen science," defining it as "scientific work undertaken by members of the general public, often in collaboration with or under the direction of professional scientists and scientific institutions." The concept of crowdsourcing data collection is not a new one. More than 2,000 years ago, ancient China used its residents to monitor migratory locust swarms. Since then many large research projects have actively involved members of the public. The practice became so common that the phrase "citizen science" was coined by Alan Irwin and Rick Bonney.[6]

I am a computational chemist interested in the structures of bioluminescent proteins, so it should come as no surprise that my first example of citizen science is a massive crowdsourced project to understand how proteins fold, folding@home. Many research groups are trying to use computers to calculate three-dimensional protein structures from their amino acid sequence. Unfortunately, this is not possible yet. The process of protein folding is still a mystery waiting to be solved. It is an important one, with many consequences for science and medicine. If proteins misfold they no longer function properly, and diseases such as mad cow disease and Alzheimer's disease can result. Protein folding normally takes about one-thousandth of a second. This is extremely fast, but the process is so complicated that even the largest supercomputer cannot solve the problem. There are hundreds of computational chemistry groups working on the protein folding problem, trying to solve it by simplifying the problem and using major computational power or machine learning, as described in chapter 7. This challenge ranks as one of the toughest problems in biology.

Folding@home is an attempt to make use of the computational capacity that is being wasted when home personal computers are sitting around not being used. Anyone can download folding@home software from the web. It is free and looks just like a screensaver to the user. As soon as the computer has been sitting idle for a predetermined time, the folding@home software kicks in and starts calculating. Any keystrokes stop the calculation, and the picture of a molecule disappears from the screen. The folding@home app runs on Linux, Windows, Apple, and Android operating systems. It is one of the world's largest distributed computing networks. As of October 2016, folding@home had a total computing power of over 100 petaflops, making it the fastest computing system in the world

Since October 2000, more than 900,000 people have downloaded the folding@home software, and it now runs on over 100,000 computers around the world every day. Buying and running 100,000 computers would cost about $50 million, and it would be a nightmare to maintain them. According to Professor Vijay Pande, who is in charge of the project and is based at Stanford University, "This is like having a whole new kind of funding agency for research—namely, the general public donating its computers. (Since this writing, Pande has left academia to become a venture capitalist using artificial intelligence in the biopharma space.) When you factor in the maintenance they are doing, the operating system upgrades, and so on, that's a gigantic resource!"[7] The program is primarily downloaded by people with interests in computers, biology, and fighting diseases, as well as teachers who find that folding@home is a unique way to get students interested in science. Using the power of distributed computing, the folding@home group has published more than 200 papers. To help compensate users for their computer use, monthly donor statistics are calculated, and foldingcoins are awarded. These are cryptocoins that can be exchanged for bitcoins on cryptocurrency exchanges. Poem@home and Rosetta@home are two distributed computing networks that work just like folding@home, both also focused on calculating protein structures.

"Traditional forms of large-scale computing—building your own cluster, buying time on a supercomputer, buying time on commercial clouds—are all expensive. Many scientists can't afford large-scale computing," says David Anderson, a research scientist at UC Berkeley's Space Science Lab and director of SETI@home. "Volunteer computing tries to solve this problem."[8] This is why there are dozens of distributed computing networks hoping to use your computer's downtime more effectively. Most are addressing chemistry and astronomy problems. SETI@home's home page describes it as "a scientific experiment, based at UC Berkeley, that uses Internet-connected computers in the Search for Extraterrestrial Intelligence (SETI). You can participate by running a free program that downloads and analyzes radio telescope data." The SETI@home app searches through data collected from the Arecibo radio telescope in Puerto Rico and the Green Bank Telescope in West Virginia to find radio transmissions that might indicate the existence of extraterrestrial intelligence.

Although distributed computed systems certainly provide an important service to science, it could easily be argued that the computer owners are funding the research rather than doing it, that they are science supporters rather than amateur scientists.

The Global Biodiversity Information Facility (GBIF) and iNaturalist rely on traditional amateur scientists and are perhaps better examples of citizen science in action. iNaturalist.org was founded in 2008. The premise of the site is that citizen scientists take photos of plants and animals, which they post with their locations and observations. Other naturalists and scientists on the site identify the species and can use the information to monitor changes in biodiversity. iNaturalist.org has also used the vast amounts of photos and information gathered from its citizen scientists to train an artificial neural network to identify the species of the organisms in most of the animal/plant pictures. In June 2017 the site released an app that uses an artificial intelligence algorithm to identify the species of plant or bird photographed.[9] Many of the iNaturalist.org postings are deposited in the GBIF, where they are part of a database of hundreds of millions of "species occurrence" records. Half the observations come from citizen scientists. In its own words, the GBIF "is an international network and research infrastructure funded by the world's governments and aimed at providing anyone, anywhere, open access to data about all types of life on Earth."[10] The facility estimates that its database has been used for more than 2,500 peer-reviewed papers in the last 10 years.[11] The GBIF and iNaturalist.org use the large number of citizen science postings to give us a global picture of what's happening to our biodiversity, while at the same time educating us and enticing us to participate in the protection of the planet's biodiversity.

There are some concerns with using citizen science. The animal sightings and geospatial information sent to sites like iNaturalist.org could be used by poachers to find rare and elusive wildlife. And there are no restrictions on the way health monitoring apps such as PatientsLikeMe use the medical data they collect.

SCIENTIFIC ART

Eduardo Kac (pronounced cats) is an example of a new breed of amateur scientist. He is the first and by far the most famous of the transgenic artists (artists who use cross-species genetic modifications in their art, genetically modifying living organisms so that they make proteins normally only found in other species). Kac was born in 1962 in Rio de Janeiro. During the 1980s he protested against the Brazilian dictatorship by giving "performance art" demonstrations on Ipanema Beach, reciting porn poems while wearing a pink miniskirt. He is currently a professor in the Art and Technology Department at the School of the Art Institute of Chicago. Although he has taken some bioengineering workshops, he has a PhD in art and never studied biochemistry, molecular biology, or chemistry.

Chapter 5 introduces green fluorescent protein (GFP). By tagging a protein with a fluorescent protein, one can see where and when a protein is made in a living organism. Most of my research is focused on fluorescent proteins, which have changed the way science is done, fascinating transgenic artists and DIY scientists along the way.

Kac produced two exhibits based on GFP technology, *GFP Bunny* and *The Eighth Day*, both part of his *Creation Trilogy*. While the first part of the trilogy, *Genesis*, didn't involve GFP, it is worth describing due to its relationship with the other two pieces and because it is thought-provoking.

In *Genesis* Kac translated Genesis 1:26, "Let man have dominion over the fish of the sea and over the fowl of the air and over everything living that lives upon the Earth," into Morse code. Since both DNA and the Morse code are made up of four different characters, Kac was able to convert the dots, dashes, and spaces between letters and words in the Morse coded version of this passage into the DNA nucleic bases C, T, G, and A, respectively. He then hired a biotech company to synthesize the "Genesis gene," which was injected into fluorescent bacteria. The gene was artificial and was probably not expressed in the bacteria. Visitors to the exhibit saw a projected view of the bacteria if, and only if, they switched on a UV lamp that briefly irradiated the bacteria and mutated them, thereby rewriting Genesis 1:26. The exhibit was shown at galleries in Linz, São Paulo, Chicago, New York, Yokohama, Athens, Madrid, and Pittsburgh. For each show, a new "Genesis gene" was created. In some cases a web link to the exhibit was established, and web surfers were given the opportunity to view and thereby mutate the bacteria. The Genesis exhibit premiered at Ars Electronica in 1999, and after the show Kac took the mutated bacteria back to the lab and had the modified "Genesis gene" sequenced, converted to Morse code, and translated back into English. Most of the mutations were nonsense mutations, but some made sense and were interesting; for example, "fowl" was mutated to "foul." In *Genesis*, Kac tried to break the barriers between art and life. It is important to note that this was done in the 1990s, when sequencing was expensive, and molecular biology was in its infancy. Today the project would be fairly trivial.

Alba is a cuddly albino rabbit that hops around, snuffles its nose, and munches carrots just like any other rabbit. Turn off the lights, switch on blue lamps, and it becomes *GFP Bunny*, a transgenic artwork. Alba, Spanish for "dawn," is both alien and cuddly. She changes from lovable family pet to a disconcerting vision of the future, a science fiction pet with an eerie green glow emanating from every cell, from her paws and especially her eyes.

Alba was created in 2000 by Louis-Marie Houdebine of the French National Institute for Argonomic Research.[12] The GFP bunny is part of the second work in the *Creation Trilogy*, but there was supposed to be more to the GFP bunny piece than just Alba. The dialogue created by the pet/alien dichotomy and the social integration of Alba were important parts of the exhibit. Alba's public debut was scheduled for an exhibition of digital art in Avignon, France. Kac and Alba were going to live in a faux living room created in the gallery, signifying how biotechnologies are entering our lives, even the privacy of our living rooms. However, on the eve of the show the director of the institute that had created Alba refused to release her to Kac. This fueled the dialogue portion of the exhibit, and soon Alba was competing with the Olympics for headlines in the *Boston Globe*, *Le Monde*, the BBC, and ABC News. *GFP Bunny* was meant to be a political project that would break down the barriers between art, science, and politics, and in this it succeeded. For many people, fears of genetically modified organisms, the human genome project, and cloning were realized when they saw photos of Alba's strangely fluorescent eyes. Kac used Alba as a symbol for all trans-genically modified organisms and of what is possible with biotechnology; she was meant to be provocative. Despite many detractors, Kac's project also had many supporters. The science fiction writer Robert Silverberg was one person of note who entered the GFP bunny debate. He wanted to know why scientists can create transgenic organisms while artists can't and whether breeding a phosphorescent (his word) rabbit is any sillier than breeding a dachshund.[13]

For *The Eighth Day*, the last part of the *Creation Trilogy*, Kac modified some amoebae, fish, mice, and plants by adding the GFP gene to their ge-nomes, then placed them in a clear, four-foot-diameter plexiglass dome, a transgenic biosphere. While most transgenic organisms have been developed in isolation, the dome in *The Eighth Day* is meant to symbolize a new ecology that is forming between genetically modified crops in the United States. It is the eighth day in the creation of man and Earth (a foreshadowing of CRISPR?). The centerpiece of the display is a robot that is driven by green fluorescent amoebae called *Dyctiostelium discoideum*. When they are active the robot goes up; when they are quiet it goes down. The robot also has a camera attached to it that can be controlled by web participants.

It took Kac and plant biologist Neil Olszewski six years to create Edunia, a petunia that has a part of Kac's immunoglobin gene expressed in its veins. According to Kac, "There's a duality here—on the one hand, it's a living thing like any other flower, it needs light and good soil, attentive watering to grow. On the other hand, the red veins in the flower carry my own DNA; I decided to give the Edunia the very same gene that in my body seeks out and rejects foreign matter."[14]

Eduardo Kac is a pioneer in the field of transgenic art, which enabled him to collaborate with interested scientists and labs. Today there are many more artists hoping to combine their art with science and scientists hoping to combine their science with art. A lot of this collaboration occurs in community labs, the scientific equivalent of maker spaces.

COMMUNITY LAB SPACES

Genspace is the first nonprofit community biotech lab established in the United States. It was started in 2009 in Brooklyn, New York. The idea for the project came from Ellen Jorgensen, who wanted a lab space that was open to everyone and would foster innovation, diversify biotechnology, and establish a space in which people could "take classes and putter around in the lab in a very open friendly atmosphere."[15] The time was right. There was a pool of disenfranchised graduate students, artists with an interest in using science as their new canvas, and highly skilled professionals with ideas and projects they couldn't pursue in their day jobs, all interested in the concept of a community lab. The 2008 recession had led to the downsizing and collapse of small biotech start-ups, which forced them to sell their equipment on eBay. Jorgensen, a molecular biologist with a PhD from New York University who has had various positions in the biotech industry, put out a call for like-minded people. They met in science journalist David Grushkin's apartment to talk about biotechnology, the need for lab space, and "to learn more about bioengineering by inserting a gene into bacteria that caused it to glow green."[16]

From that group Genspace slowly grew. In the core group that founded Genspace was Nurit Bar-Shai, an artist. She was interested in GFP and contacted me to talk about fluorescent proteins. We emailed back and forth, and I gave a few talks at the Genspace labs, which are located on an upper floor in the Metropolitan Exchange Building, a block away from the Brooklyn Academy of Music (BAM). The first time I went I got off on the wrong floor. One of the people in the building gave me a brief tour on the way to the Genspace labs. The owner of the Metropolitan Exchange, Al Attara, has attracted a variety of entrepreneurs to the building with cheap rent, communal kitchens, and a symbiotic workplace. This was the perfect location for a community biotech lab. The open-plan floors of this old bank building were packed with walls of old equipment separating groups of desks occupied by young architects, artists, and biotechnologists all bustling with energy and ideas. Al, the building's owner, is not happy with the building's name. "I want to rename it the Brooklyn Arts and Design Arena—or BADA. Since we're in the BAM District, it'll be BADA-BAM," he said in a *New York Times* article.[17] BADA-BAM would certainly capture the spirit of Genspace's energy. Most of the community lab's members are not scientists, and a lot of the energy is devoted to teaching and training members and students from local underfunded high schools. The lab only qualifies as a biosafety I lab, which means it is suitable for handling life forms that present no risk to humans.

From its very inception, Genspace and its founders have suffered from negative public misperceptions. Ellen Jorgensen recalls her first interactions with the press after forming Genspace: "The more we talked about how great it was to increase science literacy, the more they wanted to talk about us creating the next Frankenstein."[18] These fears that DIY biologists (DIYbio) or biohackers will be able to able to cause themselves or even others harm have grown, particularly since the advent and commercial distribution of CRISPR kits. In 2013, David Grushkin and Piers Millet, deputy head of the Biological Weapons Convention Implementation Support Unit of the United Nations, did a survey of DIY biologists.[19] They found that most work together, and only 8 percent of the respondents worked alone in their own home labs. Biohackers are very interested in idea sharing, open-sourcing techniques, and transparency. Community labs are sprouting up all over the country, and most cooperate with authorities to ensure that they have no accidents and that their facilities aren't abused.[20] However, biohackers have a large variety of motivations. Medical doctors and biochemists want to examine diseases that are of personal importance to them or their families; retired scientists want to continue their research; and bankers and software engineers switch careers to being transgenic artists, cyberpunks, and anarchic biohackers.

Biohacking provides a research space for high-risk projects that aren't always designed to lead to tenure or new products. Furthermore, it has allowed wannabe scientists, graduate students, transgenic artists, doctors, teachers, and industrial researchers to try out their own ideas in their own spaces. In garages and scientific maker spaces, individuals are altering their own genetic codes, building things, making cells glow, and democratizing science.

BIOHACKING

Josiah Zayner, age 38, is a biohacker interested in pushing boundaries. He has a PhD in molecular biophysics and his own company, The ODIN, which sells kits and instruments for home scientists. Zayner sees himself as a scientific adventurer and rebel researcher. He is willing to experiment on himself, to try new and experimental techniques, and to circumvent the Food and Drug Administration (FDA) in order to improve his own body. "I want to live in a world where people are genetically modifying themselves," he says.[21] In 2016, he released a YouTube video titled "How to Genetically Engineer a Human in Your Garage," showing his attempt to genetically modify himself so that cells in his arm would express GFP.[22] The self-experiments were a partial success. While Zayner was never able to observe any fluorescent skin cells when he biopsied his skin, he and an independent lab were able to show that his cells expressed some GFP in the sites he had injected with the virus. There just wasn't enough GFP for its fluorescence to be visible.

Not satisfied with the results of his attempts at fluorescent self-modification, Zayner upped the ante by trying to use CRISPR to suppress the myostatin production in his arm. Myostatin is a muscle-growth-inhibiting protein (discussed in chapter 9) that leads to double-muscled animals. Always the showman, Zayner's self-hack was performed in front of the audience of the SynBioBeta workshop held in October 2017 in San Francisco. His "experiment" was the first attempt at human CRISPR-mediated genetic modification. It was both a show and a proof-of-concept experiment. Some of Josiah Zayner's muscle cells were probably modified by his CRISPR myostatin inhibitor system, but the CRISPR delivery systems he used were not sophisticated enough to modify sufficient muscle cells to make a significant visual difference to his arm's muscle growth. However, it is just a matter of time before efficient delivery systems are available to biohackers like Zayner.

Zayner has many detractors. According to an opinion piece by Marcy Darnovsky, executive director of the Center for Genetics and Society, entitled "Hacking Your Own Genes: A Recipe for Disaster," many in the biohacking community call Zayner "a publicity-seeking stunt man, perhaps deluded by touches of toxic masculinity and techno-entrepreneurial ideology, peddling snake-oil with oozing ramifications."[23] In 2017, the FDA, without directly pointing a finger at Zayner, released the following statement: "The FDA is aware that gene therapy products intended for self-administration and 'do it yourself' kits to produce gene therapies for self-administration are being made available to the public. The sale of these products is against the law. The FDA is concerned about the safety risks involved."[24] In 2019, the California Department of Consumer Affairs reacted to Zayner's CRISPR demonstration by opening an investigation of him to establish whether he was practicing medicine without a license, and it instituted the first law in the United States to directly regulate self-administered gene therapy. Starting in January 2020, it will be illegal to sell gene-therapy kits (CRISPR kits) without a notice displayed conspicuously, "stating that the kit is not for self-administration."[25]

Bill Gates, Steve Wozniak, and Steve Jobs started the personal computing revolution in their garages. Zayner believes the biohackers and their new DIY research spaces will have a similar impact on traditional science. He would like to expand biohacking and sees himself as the spokesperson for the new movement. At Biohack the Planet 2017 conference he said, "We buy our equipment on eBay. We run it out of our garages, our kitchens, our sheds and we don't fucking have review boards. There's nobody to tell us what to do.

There's no committees who can sit there and say you can and can't do this. No, we make that choice and because of this everybody thinks we're going to destroy the world, well fucking-A the world's already destroyed and bio-hackers are the only ones who can motherfuckin save it."[26]

Zayner has some surprising supporters. George Church, a geneticist at Harvard, is an adviser to Zayner's DIY company. Furthermore, since his CRISPR demonstration Zayner has had hundreds of email requests from people interested in self-modifying themselves with CRISPR. "The barrier of possibility is broken," Zayner says, "So now the fun begins."[27]

I suspect that in the long run Zayner's public experiments will do exactly what he is rebelling against: result in new guidelines and regulations on biohacking. I think he is putting places such as Genspace in danger, lending credence to detractors who fear the making of the "next Frankenstein's monster."

Jennifer Doudna, one of the discoverers of CRISPR, has said, "The thing I worry about the most is primarily just people getting out ahead of the technology itself."[28] This may just be an example of someone outpacing CRISPR.

Predicting future trends is a risky but entertaining business, and I certainly don't fault Daniel Koshland Jr. for his prediction in 1992 that the demise of amateur science was just around the corner. He was right that there haven't been many amateur scientists making breakthroughs in the laboratory sciences. This certainly doesn't mean that amateur scientists in the lab have been marginalized. Far from it; DIY science, biohacking, and citizen science are burgeoning, are very active, and get a lot of media attention. At the same time, citizen science and crowdsourced computing have most definitively contributed to expanding our understanding of science. And while community science facilities and biohackers have democratized science, it remains to be seen whether rebel biohackers can make the transition from social media novelties to true transformers of science.

Part Two

Doing Science

Chapter Four

The Nuts and Bolts

This chapter describes how scientists know whether they can trust published results, how the peer-review system works, and where scientists get funding for their research.

PEER-REVIEWED PUBLICATIONS

In 2018, Susan Bourne, the interim dean of science at the University of Cape Town in South Africa, and I were talking about science journals. The scientific paper is the primary mechanism for both broadcasting one's own scientific results and determining what research has been done by others and how they did it. It is also a measurement used to assess a scientist's worth, whether for funding, tenure, promotion, or getting a job. I like how Susan succinctly summed up her thoughts on the scientific publishing process and probably those of most other scientists: "The system is completely crazy. The taxpayer funds the research, the scientists do the research, write it up for free, do all the editing, all the peer reviewing—the publisher gets everything for free. And then the taxpayer has to pay for us to get access to it again."

Prior to the 1600s, scientists privately communicated their findings and ideas in letters, gave public lectures, and wrote books once the experiments were all done and their ideas and theories had matured. There was no way of publishing increments of one's research. When the advent of scientific journals allowed scientists to publish chunks of their work, "scientists from that point forward became like the social insects: They made their progress steadily, as a buzzing mass."[1] At first the papers were shorter, less formal, and more readable than they are today. As the research became more specialized, papers became longer and contained more jargon.

I sometimes think of scientific papers as puzzle pieces. Nature is full of intriguing puzzles for researchers to solve. The jigsaw pieces don't come in a box, with the number of pieces listed and a picture of the solution on the lid. To solve any of nature's puzzles, researchers need to find the pieces, then try to put them in the correct place. Some puzzles are much more important than others, and within the puzzles themselves some jigsaw pieces are more central than others. Scientific research is all about finding the pieces, putting them together, and trying to extrapolate to determine the big picture even when some pieces are still missing. Some puzzles lead to new understandings, others form the basis of new theories, and yet others result in new techniques. When a puzzle reaches a certain stage, it becomes easier and easier to put in the pieces; the research accelerates. The breakthrough occurs when the pictures on the puzzle become visible, when a central piece is placed that allows whole new areas to emerge. An important puzzle can lead to the start of many other new puzzles.[2]

Each year about 1.8 million papers are published in roughly 2,800 journals.[3] The journals are not all equal. *Science*, *Nature*, and *Cell* are the most prestigious ones; the important puzzle pieces are published in them. Getting a paper published in one of these journals assures the authors of a wide readership and significant prestige, and in turn the reader knows that the papers have withstood rigorous peer review and have been judged to be of importance to all scientists. The "impact factor" of a journal (the annual average number of citations per paper published in the journal in the previous two years) is an attempt to quantify its prestige. (The 2018 impact factor for *Science* was 41.1, which means that the average paper published in *Science* in 2015 or 2016 was mentioned in 41.1 papers in 2017. The parallel impact factor for *Nature* was 41.6, almost the same.) Publishing one's work in the journal with the highest impact factor is important and something of an art. Being too ambitious in journal shopping leads to rejections and delays, while taking the safer route and submitting to a journal with a lower impact can lead to less of the needed exposure and prestige that parlay into grants, jobs, tenure, and fellowships.

Nature receives about 200 manuscripts a week but can publish no more than 8 percent of them. Upon receiving a manuscript, a staff editor with expertise in the area covered by the paper makes a first cut and within a week decides whether the paper should be sent for external review or be returned to the authors.[4]

In the next step, *Science* and *Nature*, like most other science journals, use a single-blind peer-review system to evaluate their manuscripts. The papers are sent to at least two external referees, who also have expertise in the research area covered in the paper. In the single-blind process the reviewers know who wrote the paper, but the authors never officially find out the identity of the outside experts (although journals have begun experimenting with giving reviewers the option of signing off on their reviews). Science, despite its huge expanse of subjects, can still lead to remarkably small circles of experts, especially given how highly specialized and specific certain subjects can be. As a result, it's often not out of the question that researchers could accurately guess who their reviewers are. This single-blind process creates some problems of bias against women (see chapter 2) and in favor of well-known researchers from prestigious academic institutions.[5]

Peer review is not new; it has been around for at least 350 years. Henry Oldenburg (1618–1677), the editor of *The Philosophical Transactions of the Royal Society*, may have been the first editor to use the system. Since the 1960s the number of journals has ballooned, and the need for impartial experts capable of reviewing scientific manuscripts has grown. Peer review is done on a voluntary basis; academics are not compensated for the reviews they write. The majority of scientists see this process as pivotal to scientific progress, and as such they are willing to volunteer their time to peer review. That doesn't mean, however, that they are all happy with the process as it stands. More than 200,000 academics track and verify their peer reviews on a website called Publons. In 2018, the service analyzed the million-plus reviews conducted for the 25,000 journals in its database and conducted "The Global State of Peer Review" survey of more than 11,000 researchers.[6] It found that most editors are from "leading science locations" such as the United States, Europe, and Japan, and that they tend to look for reviewers in their own backyards. Consequently, researchers from the United States, United Kingdom, and so forth write nearly two peer reviews for each manuscript they have submitted, whereas researchers from countries such as China, Brazil, India, and South Africa do about 0.6 review per submission. Reviewers from "emerging economies" are more likely to agree to review a paper. Their reviews are returned more promptly but are shorter. Only 4 percent of journal editors are from "emerging economies." It takes about five hours to write a review, which means that each year scientists spend about 68.5 million hours reviewing papers without compensation from for-profit publishers. The average review is returned in 16.5 days and is 477 words long. The unpaid cost of peer review in 2008 was estimated to be $3.5 billion.[7]

The Publons survey also showed that reviewer fatigue exists. In 2013, editors had to ask an average of 1.9 people to get one review. That number rose to 2.4 in 2017. The number of published articles has grown by 2.6 percent per year since 2013. All these papers have to be peer reviewed.

On August 26, 2016, Sarah Reisman and her students published a paper titled "A 15-Step Synthesis of (+)-Ryanodol" in *Science*.[8] Sarah had graduated in 2001 with a BA in chemistry from Connecticut College, which is where I teach. I know her and used the opportunity provided by writing this book to talk to her about the state of science today and the path of her paper from conception to citation to conversion into scientific currency.

Sarah is a full professor of chemistry at Caltech, a winner of the Arthur C. Cope Scholar Award and the Tetrahedron Young Investigator Award for Organic Synthesis, and the first recipient of the Dr. Margaret Faul Award for Women in Chemistry. She says, "I waited before submitting this paper. I felt it was a real example of synthesis that we are super proud of. I know it might be to my detriment, but every time I have a paper rejected it kills a little part of my soul. I can only get so many rejections in my career. I am very careful when and where I submit."

Sarah doesn't know who reviewed her manuscript, and while I didn't ask her to speculate, I certainly can. Her paper describes a new efficient synthesis of ryanodol, an insecticidal molecule derived from the tropical shrub *Ryania speciosa*. It can change signaling between our cells, which can affect how we move and think. Pierre Deslongchamps, the University of Sherbrooke chemistry professor who reported the first synthesis of ryanodol in 1979, says, "Today, the emphasis in total synthesis of complex natural or nonnatural products is to discover a strategy that will allow us to make the product in a minimum number of steps."[9] He made ryanodol in 41 steps. In 2014, Masayuki Inoue of the University of Tokyo reduced the synthesis to 35 steps. I bet Deslongchamps or Inoue, or even both, were asked to review Sarah Reisman's 15-step synthesis. As impartial experts in the total synthesis of ryanodol, they would have been asked by the editors of *Science* to determine whether the research described in Sarah's paper was presenting a significant major advance in the field and was of a quality commensurate with the expectations of the journal.

The peer-review system is far from perfect. Fraudulent (see chapters 10 and 12) and flawed papers are published regularly, and papers with innovative ideas are often turned down. Many famous papers were initially rejected before going on to change how whole areas of science are done. Two of them are discussed in this book. Karl Deisseroth and Edward Boyden's paper describing the first application of optogenetics was rejected by both *Science* and *Nature* (see chapter 8) before being published by *Nature Neuroscience*, and in April 2012 Virginijus Šikšnys submitted a paper to *Cell* showing that Cas9 was the enzyme responsible for cutting DNA in CRISPR. The editor did not send out the paper for review, and Jennifer Doudna beat Šikšnys to publication (see chapter 9).

About four months after submitting their paper to *Science*, Reisman and her students got to see their names on the *Science* website. Their manuscript had been accepted and was ready for printing.

I once waited nine months for a paper of mine to be reviewed. Fortunately it was accepted with minimal changes, but such long waits are unacceptable. I could have been scooped, or my undergraduate coauthors' acceptance to medical and graduate school might have been affected by this delay. In response to complaints that the review process was taking longer and longer, in 2016 *Nature* commissioned a study of the time taken for papers to move from submission to acceptance.[10] This period includes the time taken for authors to respond to the call for revisions and additions made by the external reviewers. For some journals there had indeed been an increase in waiting time; for example, the time between submission and acceptance of *Nature* papers increased from 85 days to 150 days in the decade before 2016, while over the same period the wait for *PLOS One* manuscripts rose from 37 to 125 days. However, for the preceding 30 years the median review time for all journals had remained constant at about 100 days. Although the median time had remained the same across all journals, it had increased for papers accepted by high-impact journals. This is most likely due to increased demands from external reviewers. Leslie Vosshall, a neuroscientist at the Rockefeller University in New York City, says, "We are demanding more and more unreasonable things from each other."[11] We are writing so many papers we don't have the time to find and read anybody else's work. Since the Second World War, the number of cited papers has doubled about every nine years.[12] Vosshall says we need peer-reviewed journals as "prestige filters." Without them, "How do we find the good stuff?"[13]

Ron Vale, a cell biologist at the University of California, San Francisco, compared biology papers published in *Nature, Cell*, and the *Journal of Cell Biology* in the first six months of 1984 and 2014. He found that both the number of panels in experimental figures and the average number of authors rose by a factor of two to four during that 30-year period.[14] He concludes, "More experimental data are now required for publication, and the average time required for graduate students to publish their first paper has increased and is approaching the desirable duration of PhD training." This time is often lengthened by authors "journal shopping" their manuscripts in an attempt to place their manuscripts in journals with the highest impact factors. This is a risky practice that can lead to research groups being scooped by their competitors and may lead to some of the gender inequities discussed in chapter 2.

My first papers were all submitted as hard copies, and the papers were typeset. Fortunately, those days are gone. Digital publishing has sped up and simplified both the submission and publication processes. From 2000 to today, the median time from acceptance to publication has dropped from 50 to 25 days.

Some publishers are trying to speed up and clean up the publication process by using artificial intelligence programs to screen submitted papers for plagiarism, select reviewers, summarize the content of the papers, check statistics, and ensure that the manuscripts adhere to the journal's guidelines. According to Neil Christensen, the sales director of one of these programs, UNSILO, "The tool's not making a decision, it's just saying: 'Here are some things that stand out when comparing this manuscript with everything that's been published before. You be the judge.'"[15]

Sarah Reisman is at Caltech, where she has excellent students, but she has to compete with the superstars of the chemistry world to get these students to work in her lab. She thinks the greatest benefit of her *Science* paper was that it raised her profile among prospective graduate students: "Incoming graduate students are not very sophisticated yet and put a lot of weight on where you publish."

The fact that a paper has been published in a highly selective journal is a good indicator that the work described is legitimate and has been considered significant by editors and peer reviewers. Another measure of a paper's importance is how often other articles acknowledge the paper as the source of ideas, techniques, or relevant information. These acknowledgments come in the form of citations. The Web of Science is a searchable database and one of the most popular citation indexes, containing information from more than 90 million publications. According to a 2014 article in *Nature*, "If you were to print out just the first page of every item indexed in Web of Science, the stack of paper would reach almost to the top of Mt Kilimanjaro. Only the top metre and a half of that stack would have received 1,000 citations or more."[16]

The number of citations a paper gets is only an approximation of its worth. The truly important papers, like those on Watson and Crick's discovery of the DNA double helix and Einstein's special theory of relativity, aren't cited that often because once they have become part of textbooks, they are accepted as part of general scientific knowledge. Instead, the most commonly cited papers tend to be those that describe useful methods or techniques used by many research groups. Furthermore, citations vary across disciplines because, for example, biology cites much more often than physics.

To me, the most astounding finding of the research was that more than 25 million papers have never been cited in any other papers. Think of all the effort, time, and money that has gone into doing the research and then writing up the results. Did anyone even read these papers? Were they written to disseminate results or just to pad someone's résumé? The situation appears even worse when one considers that the Web of Science tries to avoid indexing predatory journals, journals that charge a substantial publishing fee and have little to no peer review.

I discuss unethical journals in chapter 10 of this book. There are tens, maybe hundreds of thousands more papers published every year that haven't even been peer reviewed and aren't considered in this research because they weren't indexed. Librarian Jeffrey Beall has published a list of known predatory journals on his website, https://beallslist.weebly.com/.

The scientific publishing business is one of the most lucrative industries in the world, with total global revenue of over $25 billion. Five for-profit companies dominate, publishing more than 50 percent of all journals. The largest, Reed/Elsevier, has 24 percent of the scientific journal market. In 2012 and 2013, Elsevier had profit margins of more than 40 percent, higher than Apple, Google, and Amazon. "A 2005 Deutsche Bank report referred to it as a 'bizarre' 'triple-pay' system, in which 'the state funds most research, pays the salaries of most of those checking the quality of research, and then buys most of the published product.'"[17] Not only do scientists not get paid for publishing in journals or for doing peer reviewing, even more insulting, a couple of times I have been charged to use my own graphs and figures, because I had published them in one journal and was now publishing them in another.

The internet has not been kind to most publishing companies, but the publishers of scientific journals have thrived. In fact, the *Financial Times* labeled Elsevier the "business the internet couldn't kill." The reason for Elsevier's success is a master plan it developed in 1998 to deal with the internet. Elsevier decided to bundle hundreds of journals and offer online access to the packages at a set fee each year so that any student or professor could download any journal through Elsevier's website—similar to the cable TV industry. Prior to the "Big Deal," university libraries were steadily canceling subscriptions to less-used journals. Because all the journals were bundled together, universities across the globe reluctantly signed on, as no university could afford to be without certain journals in the bundle. The internet arrived, and scientific publishing grew and grew. However, that might be changing. Starting in late 2016, universities in Germany, Sweden, and Hungary canceled their Elsevier subscriptions, and in March 2019 the University of California canceled its $11 million annual subscription to Elsevier's journals.

The peer-reviewed articles at the center of this tussle have a dual function: they are both the distributors of trusted scientific communications and an instrument for evaluating researchers. In the previously discussed 2018 Publons peer-review survey of 11,000 researchers, the following four achievements were cited as being most important to career success: getting published in respected journals (53.4 percent); securing grant funding (35.9 percent); being highly cited in respected journals (27 percent); and general research, teaching, or administrative work (18.8 percent).[18]

Sarah Reisman is concerned that scientists are changing the research they do so that it will be accepted by high-impact journals like *Science* and *Nature*. These are for-profit journals, motivated to publish controversial papers that draw more readers. Reisman says, "On the one hand it's kind of nice to see people frame the science within a broader context, but I think it also results in overselling and pitching results way above what they actually do. And that's where you capture unsophisticated readers who read the research without discerning what's going on." Overall, this system induces researchers to include an eye-catching and "sexy" result to ensure that their research appears unique, even if it has little to do with the actual goals of the research.

In September 2019, Sarah published another high-profile paper, this time in *Nature*.

IS THE SCIENTIFIC PUBLICATION SYSTEM BROKEN?

The proliferation of for-profit journals has resulted in 1.5 million papers being published annually, placing significant strain on the already far-from-perfect peer-review system, which, as we have already established, is biased against women due to the single-blind aspect. There are so many papers published that many are never read in their entirety, and most aren't cited even once. Access to journal articles is very expensive and creates serious inequities between universities and countries with differing budgets. Predatory journals that have minimal peer review and charge a substantial fee have been established to provide outlets for subpar research and pseudoscience. In his final editorial as communications officer for the British Society for Developmental Biology, Andreas Prokop writes, "I sometimes feel that we were better informed in preinternet times when information was less abundant but focused on the essentials, and when dissemination was easier because the readership was hungry for information. . . . Important information can no longer be disseminated effectively and reliably, and our community no longer has the means to develop a common voice. This clearly weakens us in times where the need for communicating the importance of fundamental science is perhaps greater than ever."[19]

My colleague in the Connecticut College Chemistry Department, Tanya Schneider, once told me how her experience with journals has changed since the advent of the internet, and her comments resonated with me. She said that today it's easy to locate and read the articles that focus on our research, whereas in preinternet days we used to page through relevant journals in their entirety more often. We were probably better-informed scientists when as graduate students and post-docs we actually sat down with a journal or two over lunch. Or maybe we just had less grading and fewer committee meetings then!

IRREPRODUCIBILITY

It seems that Goodhart's Law, which states that "when a measure becomes a target, it ceases to be a good measure," is in effect. The emphasis on the number of publications, impact factors, and citations as measures of research prowess has led to their manipulation, resulting in more papers, lower-quality manuscripts, and irreproducible results. In *Rigor Mortis: How Sloppy Science Creates Worthless Cures, Crushes Hope, and Wastes Billions*, Richard F. Harris writes that the research system has been set up for failure. The whole system is built on splashy results in high-impact journals; they are the currency used to obtain funding, tenure, promotions, and job offers. The rewards for being first and getting a *Nature*, *Cell*, or *Science* paper are substantial, and the penalties for getting it wrong or being irreproducible are minor.[20] This has resulted in many highly cited papers being irreproducible. Before launching a series of expensive studies looking for new drugs based on results from published landmark papers, the pharmaceutical company Amgen—having been burned before—examined the reproducibility of results reported in 53 highly cited publications. Forty-seven of the studies could not be repeated even when the Amgen researchers enlisted the help of the original authors. In order to obtain the researchers' cooperation, Amgen had signed confidentiality agreements with the authors, so the identity of the 47 papers remains unknown. One of the irreproducible papers had been cited more than 2,000 times, and it is impossible to know how many of those papers are flawed by their reliance on the original paper. A Bayer study found similar rates of irreproducibility in publications from high-impact journals. While at Merck, George Robertson also found many academic studies that did not hold up. "It drives people in industry crazy. Why are we seeing a collapse of the pharma and biotech industries? One possibility is that academia is not providing accurate findings," he has said. None of the companies alleged fraud, but they are all concerned that the failures are a "product of a skewed system of incentives that has academics cutting corners to further their careers."[21] A spring 2016 *Nature* survey of 1,500 scientists found that more than 70 percent had encountered problems with the irreproducibility of one or more papers. This problem is not limited to academic research; federal agencies use similar promotional incentives and have similar problems with irreproducible results and scientific misconduct. For example, in his dealings with the lead-contaminated water in Flint, Michigan, Mark Edwards of the Department of Civil and Environmental Engineering at Virginia Tech has documented cases of "scientifically indefensible" reports by the CDC and EPA. Edwards and Siddhartha Roy have also written a paper, "Maintaining Scientific Integrity in a Climate of Perverse Incentives and Hypercompetition," in which they argue: "If a critical mass of scientists become untrustworthy, a tipping point is possible in which the scientific enterprise itself becomes inherently corrupt and public trust is lost, risking a new dark age with devastating consequences to humanity. Academia and federal agencies should better support science as a public good, and incentivize altruistic and ethical outcomes, while de-emphasizing output."[22] Unfortunately, it is impossible to imagine a situation in which peer review includes an irreproducibility test. As a result, it's difficult to imagine how the irreproducibility problem can be fixed while maintaining the peer-review system as it currently exists.

HOW CAN SCIENTIFIC PUBLISHING BE FIXED?

This section discusses some solutions to the problems, ranging from simple philosophical changes, to concerted structural alternatives to for-profit publication, to an illegal but commonly used solution.

Alan Finkel, Australia's Chief Scientist (impressive title, isn't it?), has an interesting idea that would cut down on the number of publications: "Let's contemplate a rule that you can only list a maximum of five papers for any given year when applying for grants or promotions."[23] I wonder what materials scientist Akihisa Inoue, former president of Tohoku University in Japan, thinks of that. One of 256 hyperprolific authors—those who have published at least 72 publications per year—Inoue holds the record for the most publications, with his name appearing on 2,566 full papers. For 12 calendar years between 2000 and 2016, he published at least 1 paper every 5 days.[24]

Scientists are evaluated by the journals they publish in, not the quality of research they publish. How about a completely transparent review system, in which the reviews and reviewer identities are published, and in which the goal of the peer reviews is to improve the research rather than to just evaluate it, followed by a public postpublication evaluation that will reward thorough, reproducible papers and devalue work that has been overhyped and hasn't lived up to its promise?[25] The journal *eLife* has already taken the first step, experimenting with combining collaborative peer review with publishing the reviews themselves and listing the reviewers' names with each paper.

An attempt to change the publishing practices of the top researchers who publish in the elite journals, *eLife* was started and funded in 2011 by three of the largest research foundations in the world: the Howard Hughes Medical Institute, the Max Planck Society, and the Wellcome Trust. *eLife* and *PLOS ONE* (founded in 2002) are two of the largest and earliest journals to become open-access journals. In the open-access model, the journal costs are not carried by the reader; the papers are freely available online. Typically, the authors are required to pay a fee upon the article's acceptance. The *eLife* fee is $2,500, but it is waived for authors with insufficient funds. Since its inception, *PLOS ONE* has published more than 26,000 papers, but its editor, Michael Eisen, is still frustrated: "I didn't expect publishers to give up their profits, but my frustration lies primarily with leaders of the science community for not recognizing that open access is a perfectly viable way to do publishing."[26]

Some journals have adapted to the pressure from open-access journals and have adopted a hybrid model in which authors can pay for their papers to be open access, but this is not enough. In September 2018, 11 European national funding agencies released a commitment, plan-S, to only fund research that is published in open-access journals. Since then 4 more agencies, including the Bill and Melinda Gates Foundation, have joined. Eisen thinks this is "a huge moment in the fight for open access because it represents the funders finally stepping up to the plate and doing what is necessary to shift the system."[27]

Preprint servers solve some publishing woes by allowing researchers to post manuscripts online before submission and invite informal external review, which mainly occurs offline by email and Twitter. Most preprint servers also host postpublication evaluations. Mathematicians, computer scientists, and physicists have been using preprint servers for decades. ArXiv, pronounced "archive" (the "X" = Greek "chi") was opened on August 14, 1991. By the end of 2014 it had more than a million preprints. Its operating costs are about $800,000 a year, or approximately $10 per article. Preprints, early versions of manuscripts posted before peer review, are slowly gaining popularity in the biological sciences, particularly after being endorsed by some major funding agencies and Nobel laureates. Cold Spring Harbor Laboratory runs the most popular preprint server in the life sciences, biorXiv, which was launched as a nonprofit service in 2013.[28] The preprints on biorXiv are published within 24 hours of receipt and given a digital object identifier (DOI), revisions are time-stamped, and anyone can read and comment on a paper. "The minute a research story gets into the public domain, it benefits from the collective power of different brains looking at a problem," says Ron Vale.[29]

Frustrated after repeatedly being denied access to articles she wanted to read, Alexandra Elbakyan attempted to launch a Russian-language open-access journal in 2011. It was to be similar to *PLOS ONE*. She failed. But in the process she found a new way to take on the big publishing companies. She started and still runs Sci-Hub (https://sci-hub.tw/ and https://whereissci-hub.now.sh/), a network of servers that store and index 70 million papers, two-thirds of all published research. Anybody can access the servers and download the papers for free. Compare this to the Elsevier average charge of $31.50 per paper. Nobody knows where Elbakyan gets her papers from, and she is not telling. It is commonly assumed that she has some collaborators who have found a way to download complete library subscriptions from academic libraries and regularly update the Sci-Hub server. One can see why Elsevier is peeved, has sued Elbakyan, and is trying to shut down her servers. Elbakyan believes her fight against the publishing behemoths is a righteous one. On the Sci-Hub home page she says, "We fight inequality in knowledge access across the world. The scientific knowledge should be available for every person regardless of their income, social status, geographical location and etc. Our mission is to remove any barrier which is impeding the widest possible distribution of knowledge in human society!" It is her belief that Sci-Hub is the only way researchers in poorer countries can get access to published research. She is correct; only a quarter of the downloads from the Sci-Hub server are from researchers working in the 34 richest countries. The remaining three-quarters of Sci-Hub's users are from poorer countries, and they probably have no alternative source for papers written by scientists with the express purpose of being accessible to other scientists. [30]

RESEARCH FUNDING

The road to becoming a scientist with your own research lab is very similar throughout the world. The most important step is a three- to eight-year PhD doing a research project in an established researcher's lab. The length of the program depends mainly on the discipline and country. Doctoral programs in physical sciences tend to be shorter than those in the biological sciences, and European and UK graduate degrees take less time than North American ones. Graduate students in the sciences are paid around $30,000 per year in the United States. Doctoral programs can be stressful and demanding; 50 percent of graduate students leave without completing their PhDs, and approximately one-third of graduate students have developed or are at risk of developing a psychiatric disorder such as depression. [31] Newly minted PhDs can enter industry, but in order to run their own research labs in academia or research institutes, they need to do a two- to four-year post-doc. They still have quite a mountain to climb. While the main aim of the PhD is to do a research project and learn how to do research and write it up, the aim of a post-doc is broader, often to oversee labs or sections of labs. Based on their publications (impact factor, number, and citations), academic pedigree, letters of recommendation (especially from PhD and post-doctoral mentors), and grant writing, potential post-docs enter the job market looking for research positions. Fifty-five percent of graduate students are contemplating academic careers. However, only 3.5 percent of PhDs get a permanent academic post, and just 0.5 percent will become full professors. [32]

Scientific research is funded by taxpayer dollars, private foundations, and industry. It can be divided into three categories: basic, applied, and developmental. The NSF defines applied research as "aimed at solving a specific problem or meeting a specific commercial objective" while basic research is defined as "activity aimed at acquiring new knowledge or understanding without specific immediate commercial application or use." Basic and applied research each comprise one-sixth of all research funding, while developmental research makes up the remaining two-thirds. Industry funds nearly all developmental research, which is focused on converting scientific ideas into products.

Twenty-six US government agencies, including the NSF, NIH, and Department of Energy (DOE), fund a significant portion of the research conducted. The federal R&D as a share of the country's gross domestic product (GDP) declined to 0.78 percent in 2014 from about 2 percent in the 1960s.

The NIH funds most of the medical and life sciences research in the United States; the funding rates for its grants fell from 30.5 percent in 1997 to 18 percent in 2014, and the average age of first-time awardees has climbed to 43 years. It is getting harder to get funding for research.

In 2013, federal funding for basic research fell below 50 percent for the first time since the Second World War. In the 1960s the federal share of funding basic research was 70 percent; it dropped to 44 percent by 2015. In the same period the spending ratio went from 2 to 1 in favor of federal funding to the current 3 to 1 in favor of industry funding.[33] There is no way funders can predict which basic science is going to lead to breakthroughs in science. Who could have foreseen that catching over 1 million jellyfish to establish the mechanism of their bioluminescence (fluorescent proteins; chapter 5), determining the function of clusters of repeating palindromic stretches of DNA (CRISPR; chapter 9), and identifying the mechanism by which algae orient themselves toward light sources (optogenetics; chapter 8) could change the way science is done. If the United States wants to continue being a scientific research leader, it has to carry on funding research, especially basic research.

Government agencies are restricted in how they can evaluate proposals and distribute taxpayer dollars. They tend to fund researchers who are doing the kind of research that they have already proven, by their publication record, they can do. Innovative new work won't necessarily get funded. A small survey of 400 scientists revealed that many were writing grants for work they thought would be funded, not for research they were truly interested in.[34] Grants and papers are driving science, not the other way around. This is frustrating many scientists, who became scientists to positively impact the world around them. A LinkedIn survey showed that 41 percent of scientists chose their jobs for reasons other than money or status, which is somewhat higher than the 37 percent average for all professions.[35]

Philanthropic foundations have much more freedom. The Gates Foundation and some European research foundations have been experimenting with anonymous short proposals, in which the reviewers only judge the viability and validity of the research project. This allows young investigators to compete with established ones, enables researchers to propose projects outside their immediate field of past work, and eliminates biases.

Crowdsource funding has emerged as a new niche funding source. After the initial successes of Kickstarter and IndieGoGo, science-specific crowdsourcing apps like Petridish, Walacea, and Experiment have taken over. Crowdfunding is best used to fund small-scale projects; initial experiments to see if they work and are worth writing a larger, much longer government grant; and attempts to convert one's research results into an often quirky product (e.g., glass dragons filled with bioluminescent dinoflagellates or anti-mosquito patches). The advantages of crowdsourcing are that it doesn't require many preliminary results, is quick, doesn't have to adhere to government regulations, and requires no long annual reports. Disadvantages are that there is no accounting, and the funds can be misused. Further, there are no regulations to check the validity of the results and the safety of the participants, particularly if the research is being conducted in a hacker space. Crowdsourcing will never replace traditional research funding sources; it is just a new funding source that requires the popularization of one's research, because a crowd is needed. [36]

A distressing consequence of the facts that modern science is done with large interdisciplinary collaborations and funding is distributed to researchers with proven track records is that science is being concentrated, inequalities are getting larger, and diversity is decreasing. Boston, San Francisco, and the North Carolina Research Triangle have become the Silicon Valleys of scientific research. They are being awarded billions of dollars' worth of grants, matched with equivalent amounts of venture capital funding. Science funding, research talent, and start-up companies have concentrated on the East and West Coasts, diverting scientific opportunities away from middle America, state universities, and HBCUs. Half of all NIH funding is awarded to just five states, and 20 percent of universities are responsible for 60 percent of start-ups. Compare Boston and Washington, DC, remembering that DC is no slouch: Boston has 151 percent more NIH funding and 2,010 percent more venture capital investment than DC. These imbalances between two cities at the top of the scientific food chain highlight the growing inequality and declining diversity seen across the world, not just in America. [37]

After reading about scientists and the nuts and bolts of science, you might have the impression that science is not in a good place. This is not quite true. Science is a massive, hugely complex undertaking that is growing faster and faster. Not everything is working perfectly, and there is much room for improvement. Reading about everyday science that is working well is not particularly exciting. A book like this necessarily focuses on the spectacularly good and bad aspects. Next I change focus and highlight some of the fantastic science being done all over the world.

Part Three

Old Science

Chapter Five

Recognizing a Breakthrough in Science

What Can We Learn from the Past?

Science and everyday life cannot and should not be separated.—Rosalind Franklin

Three breakthroughs from the past—the determination of the double helical structure of DNA, the chlorination of water, and the use of a jellyfish to light up medicine—introduce modern science, scientific breakthroughs, and the importance of personality in science.

It is difficult to predict whether a scientific discovery will be significant enough to warrant a Nobel Prize, to change the way science is done or understood, or even to change the lives of a significant portion of the general population. Before introducing CRISPR, optogenetics, deep learning, and gravitational waves—four recent scientific advances that in my opinion will change the world we live in—I discuss scientific breakthroughs from the past that provide context for the current state of science.

THE DETERMINATION OF
THE DOUBLE HELICAL STRUCTURE OF DNA

In 1953, James Watson and Francis Crick published "A Structure for Deoxyribose Nucleic Acid" in *Nature*. The succinct one-page article described an alpha helical DNA structure formed by two strands of alternating phosphates and deoxyribose held together by hydrogen-bonded adenine (A) and thymine (T), and cytosine (C) and guanine (G) base pairs that form the iconic twisted ladder structure. The article concluded, "It has not escaped our notice that the specific pairing we have postulated immediately suggested a possible copy-

61

ing mechanism for the genetic material."[1] The paper was an instant hit; it was immediately apparent that this was a major scientific breakthrough. This short paper was the critical jigsaw piece in an important puzzle that had intrigued many of the most important scientists of the time. Watson and Crick didn't do a single experiment, but they built a model, found the right puzzle pieces, and put them together correctly.

The first piece of this puzzle was found in pus-coated bandages in Switzerland. In 1869, Friedrich Miescher reported finding a new substance inside the nuclei of white blood cells (hence the pus-soaked bandages). He found that the substance was resistant to protein degradation and had an unusually high phosphorus content. He called it "nuclein." We now call it deoxyribonucleic acid, or DNA. Although he didn't know it, Miescher had found the genetic material of the body, the basis of all inheritance.

For some reason, Miescher's puzzle piece lay fallow. Others continued to work with nuclein, but more than 100 years passed before scientists appreciated the importance of Miescher's finding and acknowledged that he had discovered the molecular recipe book for all living things. I am sure Miescher would have been stunned to know that if all the DNA in a human cell were unraveled to form a single microscopic strand of genetic material, it would be roughly six feet long. And since we have about 100 trillion cells in our bodies, that means the combined DNA from all the cells in our body could stretch from the earth to the sun and back 49 times.

Phoebus Levene was born the same year Miescher reported the discovery of nuclein. A biochemist, while at the Rockefeller Institute of Medical Research in New York City he published more than 700 research papers (clearly a prolific puzzler). Some discoveries were critical pieces of our DNA puzzle: he discovered that the components of DNA were phosphates, sugars, and nucleic bases; that they are linked in phosphate-sugar-base units; and that the sugar was a deoxyribose in DNA and ribose in RNA. However, starting in 1910, Levene also published a number of papers proposing that DNA was composed of equal amounts of the four bases adenine (A), cytosine (C), guanine (G), and thymine (T). This was a mistake! Try to make a puzzle when there are some incorrect jigsaw pieces; you'll find it is not easy.

Fortunately, Erwin Chargaff rectified the situation in a 1950 article in *Nature*[2] in which he reported that in DNA the amounts of cytosine and guanine are always equal, and the amounts of adenine and thymine are also always equal. In addition, he showed that the relative amounts of bases found in DNA are organism specific: for example, in humans A = 29.3% and T = 30.0%, G = 20.7%, and C = 20.0%, while in the octopus A = 33.2% and T = 31.6%, G = 17.6%, and C = 17.6%. These are known as Chargaff's rules.

Most of the puzzle pieces were ready and waiting for Watson and Crick; however, they were missing the most important piece. The critical piece was not in circulation because it had not been published. It was a crisp, clear X-ray diffraction photograph of a DNA crystal taken by Rosalind Franklin. Because this part of the story is a little sordid and controversial, I make a brief diversion into Rosalind Franklin's life before addressing the photograph.

Rosalind Franklin was born into an affluent and influential British Jewish family on July 25, 1920, in London. For her PhD research at Cambridge she studied the pores in coal, research that was useful in quantifying the fuel content of coal. Her thesis was titled "The Physical Chemistry of Solid Organic Colloids with Special Reference to Coal." After the war, during a post-doc in Paris, she examined the structure of coal using X-ray crystallography.

In 1951, Franklin started a three-year fellowship at King's College, Cambridge. Although she was appointed to conduct X-ray crystallographic research on proteins and lipids, her supervisor, John Randall, asked her to focus her attention on DNA fibers instead. He did this without informing Maurice Wilkins, another chemist working under Randall at King's College who had already taken some X-ray photographs of DNA. Wilkins was peeved that Franklin was infringing on his area of research and let it show. His rancor and the fact that, according to Franklin, "at King's there are neither Jews nor foreigners," made her very unhappy with her fellowship at King's.[3]

Returning to the photograph, in 1950 Rudolf Signer, a professor of organic chemistry, brought 15 grams of highly purified DNA, which he had extracted from the thymus glands of calves, to England. He gave some of the DNA to Maurice Wilkins. But John Randall also asked Rosalind Franklin to take some X-ray diffraction photos of Signer's DNA fibers. Her photos, particularly one named "photograph #51," were much clearer and more defined than those taken by Wilkins.

James Watson first saw the photograph in late January 1953. As he later wrote in his best-selling book *The Double Helix*, "The instant I saw the picture my mouth fell open and my pulse began to race."[4] The distinctive X-shaped diffraction pattern of the photograph was indicative of a double helix structure. Watson was at King's College visiting Francis Crick's friend Maurice Wilkins when he first saw it. Wilkins, who as we know was not on good terms with Franklin, told Watson about the photograph and then went to Franklin's desk, took photograph #51 out of the drawer, and showed it to him. Two weeks later, Watson and Crick built their now-famous model of the DNA double helix using angles and spacings derived from Franklin's photograph. They had found the missing puzzle piece, in Franklin's drawer. "Rosy, of course, did not directly give us her data. For that matter, no one at King's realized they were in our hands," Watson wrote in *The Double Helix*. They interpreted Franklin's data and published the results in *Nature*, simultaneously describing the structure of DNA and its copying mechanism.

Crick did not know how to acknowledge the Franklin photo and felt he was in a very difficult position. The more he credited Franklin the more Wilkins, his friend, would feel slighted. Ultimately, Crick offered Wilkins coauthorship of the paper together with Franklin. Wilkins declined for both of them. Perhaps he didn't earn the coauthorship, but it was quite a self-centered (and perhaps sexist) action to assert that Franklin shouldn't be a coauthor either. Wilkins also asked his friend Crick to remove a reference to Franklin's "very beautiful" X-ray photos in the acknowledgments. Ultimately the acknowledgments in the *Nature* paper stated:

> We are much indebted to Dr. Jerry Donohue for constant advice and criticism, especially on interatomic distances. We have also been stimulated by a knowledge of the general nature of the unpublished experimental results and ideas of Dr. M. H. F. Wilkins, Dr. R. E. Franklin and their co-workers at King's College, London. One of us (J. D. W.) has been aided by a fellowship from the National Foundation for Infantile Paralysis.

In no way does this acknowledge the crucial role played by Franklin's photograph #51. In a footnote in an article published in 1954, a year later, Watson and Crick forthrightly admitted:

> We are most heavily indebted in this respect to the King's College Group, and we wish to point out that without this data the formulation of our structure would have been most unlikely, if not impossible. We should at the same time mention that the *details* of their X-ray photographs were not known to us, and that the formulation of the structure was largely the result of extensive model building in which the main effort was to find any structure which was stereochemically feasible. [5]

Rosalind Franklin died in April 1958, at the age of thirty-seven, from ovarian cancer. Watson, Crick, and Wilkins were awarded the Nobel Prize in chemistry in 1962 "for their discoveries concerning the molecular structure of nucleic acids and its significance for information transfer in living material." The Nobel Prize cannot be awarded posthumously, and consequently Rosalind Franklin did not get a Nobel Prize. At the award ceremony neither Watson nor Crick mentioned Franklin's role in their discoveries during their Nobel speeches. Not cool. [6]

LESSONS FROM THE DNA STORY

How does the Watson and Crick breakthrough help us find and analyze scientific discoveries made today? Here are some thoughts.

There are probably a few Mieschers out there who have made their discoveries too early for science to recognize their import. Obviously I have no chance of finding them.

Newton was right when he said, "If I have seen further it is by standing on the shoulders of giants." There is no way that Watson and Crick could have determined the helical twist and base pairing of DNA without the prior work of Miescher, Levene, Chargaff, Franklin, and others. Today more than ever we find that science relies on the work of others to supply the context and foundation for research (there are always puzzle pieces already put out on the table). A good scientist cannot work in isolation and must keep up and be familiar with the pertinent literature.

The experimental basis of the Watson and Crick paper was just one photograph, yet the paper probably should have listed four coauthors: Watson, Crick, Franklin, and maybe Wilkins. Because of the complexity and nature of modern research, multidisciplinary teams of researchers work together, leading to papers with numerous coauthors. This interdisciplinary approach leads to fantastic discoveries that could not take place without the collaborations, but it also complicates assigning credit. When the stakes are high, as they were in the determination of the DNA structure, there is likely to be a lot of competition and controversy among and within these multidisciplinary research groups.

The 1950s, when the Watson and Crick paper was published, was not an easy time to be a female scientist. Women researchers faced significant discrimination and many obstacles. In part due to the tone and condescending attitude taken by Watson toward Franklin in *The Double Helix*, many believe that gender discrimination was the reason Franklin's photograph and work were not credited. Unfortunately, a book about the state of science requires more than just a passing historical footnote about gender discrimination, because it is still alive and kicking today. Thus we have chapter 2 in this book.

THE DISINFECTION OF WATER BY CHLORINATION

In *Enlightenment Now*, Steven Pinker ranks scientific breakthroughs in terms of lives saved by each respective discovery.[7] The top three discoveries in Pinker's list are

1. the discovery of the disinfection of water by chlorination by Abel Wolman and Linn Enslow, credited with saving 177 million lives;
2. William Foege's smallpox eradication strategy, which has saved 131 million lives; and

3. the development of 40 vaccines by Maurice Hillman, among them 8 that are commonly used today (including measles, mumps, and chickenpox vaccines) and are estimated to have saved 129 million lives.

Pretty impressive! That sounds like a great way of quantifying a scientific discovery. But wait: Are they scientific discoveries? They have certainly changed our lives for the better, but have they changed the way science is done or improved our understanding of science? Not really, and that is reflected in the fact that the four scientists listed here, who together may have saved as many as 437 million lives, have not been awarded a single Nobel Prize among them.

In looking for scientific breakthroughs in the current era (see the next three chapters), I have not neglected the life-altering potential of the work.

GREEN FLUORESCENT PROTEINS

The last scientific breakthrough from the 20th century that I describe in this chapter leads right to CRISPR, gene drives, and optogenetics. I have had the good fortune to be peripherally associated with this discovery and have seen it grow.

The crystal jellyfish, *Aequorea victoria*, which drifts throughout the northern Pacific, has no brain, no anus, and no poisonous stingers. It is an unlikely candidate to ignite a revolution in biotechnology, yet on the periphery of its umbrella it has about 300 photoorgans that give off pinpricks of green light that have changed the way science is conducted. This green light allows scientists to see when and where proteins are made in living organisms; it is used in thousands of labs every day; and it has resulted in the awarding of two Nobel Prizes, one in 2008 and the other in 2014. Not only has it changed the way large swaths of science are done, but it has also led to some important lessons that are useful in finding new breakthroughs and in understanding the state of today's science.

Osamu Shimomura, a research scientist at Princeton University and Woods Hole Oceanographic Institute, spent more than 40 years trying to understand the chemistry responsible for the emission of the green light in *A. victoria*, and in the process, he caught more than a million jellyfish.

Shimomura found that *A. victoria* uses two proteins to make its green light. The first protein, which he named aequorin in honor of *Aequorea*, gives off blue light when it binds calcium in a test tube, but in the photoorgans of the jellyfish, the blue light energy produced by aequorin is entirely absorbed by a different protein, which subsequently reemits the energy as green light. This process, in which high-energy blue light is absorbed and is then immediately returned as lower-energy green light, is called "fluorescence." This is why the second protein has appropriately been named green fluorescent protein, usually abbreviated GFP.

In 1985, Douglas Prasher, a researcher at Woods Hole Oceanographic Institute, had a radical idea. He thought that it would be possible to utilize GFP's fluorescence as a tag to light up when a protein was made and show where it moved in a cell or even in a whole organism. Proteins are very small, and at that time it was impossible to see them in a cell, let alone track their movement in a living organism. By attaching GFP to a protein, one could make a modified protein that fluoresces, which would be much easier to detect, like seeing the light of a firefly even if you are too far away to see the firefly itself.

Prasher proposed taking the GFP gene and inserting it at the end of the gene of an interesting protein, just before the stop codon. In this way the protein of interest would be made with GFP tagged onto its end, and the green fluorescence would show when and where the protein was made. This is a bit like going to your favorite recipe and adding the following sentence at its end: "Now ring the bell." You have modified the recipe so that you will always know when your favorite food has been made.

At the time of Prasher's proposal, the location of the GFP gene in the jellyfish genome was unknown. Shimomura had isolated GFP from jellyfish, but he wasn't interested in its gene. Today it would be fairly easy to find the gene for a protein like GFP in a jellyfish. It would take a week or two and require just a couple of jellyfish. In the early 1980s, however, when Prasher was doing most of his research, the molecular biological techniques commonly used today were unknown, and it was an extremely difficult process.

Over a three-year period, Prasher caught roughly 70,000 jellyfish, which gave him enough messenger RNA for two rounds of gene searching. Prasher's perseverance paid off, and he finally found and sequenced the complete GFP gene. He inserted the gene into *Escherichia coli* (a bacterium), which then produced the protein. Sadly, the GFP did not fluoresce. Douglas Prasher was devastated but not greatly surprised. It seemed that GFP was not intrinsically fluorescent. Prasher gave up on the project.

Before closing shop on GFP, he published a paper in *Gene* in 1992,[8] describing how he had isolated the GFP gene and giving its sequence. A year later Martin Chalfie, a biology professor at Columbia University, found the *Gene* paper and contacted Prasher, requesting the GFP gene. Prasher sent him the gene.

Chalfie received the GFP DNA in September 1992 and entrusted a student, Ghia Euskirchen, with the GFP project. When Prasher had isolated and copied the GFP gene, he had copied 25 additional base pairs before the gene and 227 after it, which was easier than isolating just the gene itself. It was common practice at the time, and he didn't think that it would make any difference. In trying to add the GFP gene to *E. coli* so that they would express glowing GFP, Euskirchen made one major change to Prasher's procedures. She used primers that copied only the coding sequence, thus leaving out those additional base pairs, so that she could use the polymerase chain reaction (PCR), a very new technique at the time, to copy just the gene before inserting it into *E. coli*.

Two weeks after getting the GFP gene, Ghia Euskirchen was the first person in the world to see GFP expressed in *E. coli* fluoresce. In Prasher's lab, that little bit of extra DNA before and after the GFP gene was responsible for the GFP folding incorrectly in the bacteria, and it had prevented GFP from emitting its signature green glow. Euskirchen had overcome the final hurdle by amplifying the DNA with PCR and not using restriction enzymes. GFP expressed in *E. coli* was fluorescent; it didn't need an additional jellyfish protein.

Chalfie studied *C. elegans*, and he was particularly interested in how the transparent roundworm responds to touch. In 1993, his lab replaced the gene for a touch-sensitive protein with that of GFP, causing the worms to glow magnificently when touched. The paper describing the work, with Douglas Prasher listed as a coauthor, was accepted in *Science*, and the image of the genetically modified *C. elegans* associated with the manuscript was reproduced on the front cover of the magazine.

Scientists now had a new tool that would open windows to worlds barely imagined before. However, like a car standing on the top of a hill, this revolution needed a push to get it going.

Roger Tsien, professor of chemistry at the University of California, San Diego, gave GFP technology the shove it needed. Very soon after Chalfie published his paper in *Science*, Tsien expressed GFP in yeast and modified it so that it fluoresced blue, yellow, and cyan instead of green. He also created the basis of a much brighter version of GFP, called enhanced GFP (EGFP). Even now, EGFP is the most commonly used (but not necessarily the best) fluorescent protein, and there are ongoing efforts to find new and brighter-colored fluorescent proteins.

The jellyfish protein GFP is lighting up medicine and biology. Its colorful, fluorescent images have adorned many a modern science journal cover, not just because they are super cool and can be mistaken for modern art, but also because they allow scientists to extract information from living organisms that was previously hidden to them.

On October 8, 2008, the Nobel Foundation announced that the 100th Nobel Prize in chemistry was being awarded to Osamu Shimomura, Martin Chalfie, and Roger Tsien, "for the discovery and development of the green fluorescent protein, GFP." No more than three people can share a Nobel Prize. Douglas Prasher was not awarded a share of the 2008 Nobel Prize in chemistry.

A bit more about Douglas Prasher. In 1994 he gave a talk at Connecticut College. As a junior faculty member in the Chemistry Department at the college, I was in charge of the slide projector. I had never heard of GFP, which was a relatively unknown molecule at the time. In fact, only four papers on GFP had been published. More than 20 years later, I can still remember standing at the back of the lecture room, looking over no more than 15 restless undergraduates. It was not the most impressive crowd, and the tall, balding speaker struggled to hold the students' attention. He was not a natural speaker, and the importance of the material was lost on the listeners. Consequently, when Prasher stressed that GFP had the potential to be big— very big—at the conclusion of his talk, I wasn't quite convinced. However, GFP did sound interesting, so I started my academic career working on a protein that was about to become an indispensable tool in the arsenal of biomedical labs all over the world. At the time, fewer than 20 people were studying fluorescent proteins, but now more than three million experiments each year utilize these illuminating proteins to show us when and where proteins are made in living organisms, thereby increasing our knowledge of the mechanics of disease and providing new avenues in the search for their cure.

Prasher, whose seminar was the impetus for my interest in GFP, was not among the Nobel awardees. In interviews following the announcement, both Chalfie and Tsien were magnanimous and repeatedly acknowledged his contribution to their research. Martin Chalfie has often stated that he wished that there was not a three-person limit on the number of scientists who can win the Nobel Prize in each of the categories. "Cloning GFP was essential to this entire project," says Chalfie. "Without it, neither my work nor Roger's work would have been possible."[9] Tsien and Chalfie invited Prasher and his wife, Virginia, to the Nobel ceremonies and paid for their flights.

For 23 years, fluorescent proteins have been the main focus of my research. Thanks to fluorescent proteins, I have published many papers, had funding from the NIH, published three books, and attended the 2008 Nobel Awards, yet I am continuously surprised by the diversity of its uses and its importance in modern science. GFP's successors—CRISPR, gene drives, and optogenetics—are new techniques that will not only change the way science is conducted but also change the lives of our children. In my opinion the GFP story has numerous implications that are important to consider when choosing and discussing modern scientific breakthroughs.

LESSONS LEARNED FROM THE GREEN FLUORESCENT
PROTEIN RESEARCH

The 2008 Nobel Prize in chemistry rewarded both basic and applied research. While Shimomura's primary interest in GFP was its role in *Aequorea* bioluminescence, Tsien was interested in its practical applications. During the Nobel awards banquet in 2008, Roger Tsien, speaking for the three chemistry award winners, said, "We hope this prize reinforces recognition of the importance of basic science as the foundation for practical benefits to our health and economies."[10] There is very little doubt in my mind that the type of research Shimomura did, research to understand a biological phenomenon, in this case jellyfish bioluminescence, would not be funded for 20-plus years in today's research climate. The puzzle pieces Prasher needed to come up with the "fluorescent protein as a tag" idea would not have been on the table.

Roger Tsien developed brighter, faster-maturing, more photostable, and multicolored fluorescent proteins, then incorporated them into in vivo sensors. He eloquently stated the case for applied research in his Nobel speech: "Some people have at times criticized us for mainly working on techniques. I would like to draw their attention to an old Chinese proverb that says that if you give a man a fish you feed him for one day, if you teach him how to fish you feed him for a lifetime. That's why we enjoy devising fishing tackle and nets to scoop from the ocean of knowledge."[11] After their groundbreaking *Science* paper, Martin Chalfie and Ghia Euskirchen did not develop fluorescent protein probes as Tsien did, in part because Chalfie was concerned that the development of techniques would not be beneficial to his career in the biology department at Columbia University. The 2008 and 2014 Nobel Prizes in chemistry being awarded to fluorescent protein researchers shows that the tide has changed. Applied research is "in." Today, researchers who devise and refine techniques are more likely than ever before to be funded, be promoted, and have their research published in the most prestigious journals.

Luck and personality play a much larger role in the timing of scientific breakthroughs than one might expect. Had Prasher done his experiments a year later, he might have used PCR with just the coding sequence of GFP, and he would have had a fluorescent *E. coli*. There was nothing he could do about his bad luck and bad timing. This makes me wonder how many experiments have failed because the appropriate technique wasn't around when the experiments were done. An amusing law, Stigler's law of eponymy, states that no law is named after its original discoverer. In *This Idea Is Brilliant*, William Poundstone lists a series of discoveries named not after the person who found the first puzzle piece, but after puzzlers who subsequently placed important, well-timed pieces in the puzzle, for example, Halley's comet, Occam's razor, the Coriolis effect, Venn diagrams, and bechamel sauce.[12]

Timing is key. In addition to being great scientists, Chalfie and Tsien were in the right place at the right time.

Scientists also need more than their fair share of self-confidence if they want to change the way in which science is done. The scientific method requires that researchers publish their ideas and have them withstand the rigors of peer review. This is a daunting process, especially when the ideas are so radical that there is little support for them in the community. Believing in one's work and having the courage to fail are critical requirements for paradigm shifts in science. Prasher's idea to use GFP as a molecular light-bulb was an excellent one, and his experimental technique was flawless, but his timing was bad, and he lacked the drive, support, confidence, and doggedness to plow through the obstacles in his way.

Scientific reputations are partially built on conference presentations and dinner discussions, which require confidence and social skills. Unfortunately, the peer-reviewed publication and funding systems are not blind to these academic reputations, and reviewers are influenced by stature. Today the most successful scientists have outgoing personalities to supplement their intellectual prowess and scientific skills. This is an era of social media and self-promotion, and personality is playing an increasingly important role in science. The science produced by the quieter, more reserved Franklins and Prashers will find it increasingly difficult to compete (for funding, etc.) with that produced by the confident and flamboyant Watsons.

Watson and Crick, Schrödinger, and Einstein have all made theoretical breakthroughs that have changed our understanding of science. Today, science has matured to a state at which big game-changing ideas are going to become less common. As a case in point, there have been no major theoretical breakthroughs in the early 21st century. It is techniques that are changing modern science. New and improved techniques accelerate scientific research and discoveries, not only because they allow us to do our experiments faster than before, but also because they allow us to go back to results from the past and take them to new horizons using methods unavailable to the original scientists.

In the next four chapters I examine the Laser Interferometer Gravitational-Wave Observatory, deep learning, optogenetics, and CRISPR and gene drives, the newest modern scientific breakthroughs, to see what they can do and how they reflect the state of science today.

Part Four

New Science

What were the most important advances in science in the last 10 years, how do they work, how were they discovered, and perhaps most important, how are they changing science?

Chapter Six

LIGO and Virgo

Measuring Gravitational Waves
1.3 Billion Light Years Away

Up to now, humanity has been deaf to the universe. Suddenly we know how to listen. The universe has spoken and we have understood.—Professor David Blair, director of the Australian International Gravitational Research Centre

The research described in this chapter is a stunning tour de force, an amazing technical feat. It involves the largest project ever funded by the NSF, has been the basis of a Nobel Prize, and has already generated more than 600 PhD dissertations in the United States alone.

In November 1915, Albert Einstein presented a theory, the theory of general relativity, which states that space and time curve in the presence of mass, and that this curvature produces gravity. As a logical consequence of the theory (if one is clever enough to understand the theory, and I am not), one would expect a dying star to create an area in space where the gravitational pull is so strong that even light cannot escape. This makes the area invisible; it is a black hole: unseen, but "photographed" in April 2019.

Einstein's theory of general relativity also predicted the existence of gravitational waves, ripples in space-time produced by massive objects moving at extreme accelerations (e.g., two black holes orbiting around each other at increasing rates). Gravity is a very weak force, and Einstein didn't think we would ever be able to detect gravitational waves. As they pass through space, they should stretch and squeeze it by just one part in every 10^{21} parts. They are tiny. In theory (and in fact in practice), they regularly pass through us and stretch and squeeze us by the tiniest amounts, so small that we can't feel or measure the changes.

In the 1970s, Rainer Weiss at MIT in Boston and Kip Thorne at Caltech in Los Angeles were convinced they could design an instrument that would detect the stretch and squeeze of space caused by gravitational waves as they altered the distance traveled by two perpendicularly arranged laser beams. If two laser beams were to travel along two identical evacuate tubes and hit a mirror at the end of the tubes, they should return to the source at the same time and still be in phase with each other. However, upon interaction with a gravitational wave, the distance along one tube would be slightly compressed, and the distance along the other would be slightly stretched. This would put the two lasers out of phase with each other by the time they both reached the detector. It would follow that the larger the difference in phases between the two lasers, the stronger the gravitational wave.

David Kaiser, a professor of physics at MIT and coeditor of *Groovy Science: Knowledge, Innovation, and the American Counterculture*, thinks that "the idea was audacious, to say the least. To detect gravitational waves of the expected amplitude using the interference method, physicists would need to be able to distinguish distance shifts of about one part in a thousand billion billion."[1] The construction of such a sensitive instrument would be very expensive and challenging in itself, but even if it could be made, how could one distinguish gravitational waves from vibrations caused by a car crash occurring miles away, a thunderstorm, or something else? The solution was as ingenious as it was expensive: make two observatories.

Getting funding to construct the needed "observatory" was a monumental task requiring equal skills in physics, backroom negotiation, and persuasion. In 1990 Weiss, Thorne, and Ronald Drever, another Caltech physicist, persuaded the NSF to fund the construction of the Laser Interferometer Gravitational-Wave Observatory (LIGO).

LIGO consists of two 4-kilometer-long, L-shaped vacuum chambers, one located in Louisiana and the other 3,002 kilometers away in Washington State. The vacuum chambers are housed in 2-meter-tall concrete pipes that have to be raised a little less than a meter at their ends to account for the curvature of the earth. Gravitational waves that originate tens of millions of light years from Earth distort the 4-kilometer mirror spacing at the ends of the vacuum chambers by about 10^{-18} m (one-thousandth the width of a proton). To ensure these minuscule distortions aren't caused by local car crashes, earthquakes, and so forth, the two LIGO observatories were constructed thousands of kilometers apart. To be a gravitational wave, the same deflection has to be observed at both LIGO interferometers 10 milliseconds apart, the time it takes to travel 3,000 kilometers at the speed of light.

At $1.1 billion spent over 40 years, LIGO was and still is the largest project ever funded by the NSF. Rich Isaacson was a program officer at the NSF and played a significant role in getting the project funded. "It never should have been built," he told Nicola Twilley, a writer for *The New Yorker*.

"It was a couple of maniacs running around, with no signal ever having been discovered, talking about pushing vacuum technology and laser technology and materials technology and seismic isolation and feedback systems orders of magnitude beyond the current state of the art, using materials that hadn't been invented yet."[2]

But the project was funded, and ground was broken in 1994. Seven years later, LIGO was up and running. From 2001 to 2010, hundreds of scientists combed the LIGO data looking for gravitational waves. No luck. Upkeep and modification of the observatory is no trivial procedure. The equipment is so sensitive that when the researchers want to make an adjustment to the optical components, they have to put together a portable clean room, sterilize everything, and put on full-body protective suits in order to prevent anything, even an eyelash or a skin cell, from damaging the sensitive equipment. In 2010, they shut the whole shebang down for a five-year, $200 million upgrade, installing newer, better lasers; mirrors; data analysis programs; and seismic isolation technology. Every component was specially designed and on the cutting edge of technology. For example, the new and improved mirrors were the best available in the world, polished to within a hundred-millionth of an inch of a perfect sphere. At a bit more than a foot wide and weighing close to 90 pounds each, they cost $500,000 apiece.

GRAVITATIONAL WAVES ARE DETECTED

The updated LIGO was scheduled to restart its data collection on September 18, 2015. The weeks and days before the big day were hectic as everyone prepared. Anamaria Effler was an operations specialist at the Hanford site, where they were frantically working through the weekend to ensure that all the systems were up and running for the official reopening. On Sunday, September 13, she was doing some last-minute tests. "We yelled, we vibrated things with shakers, we tapped on things, we introduced magnetic radiation, we did all kinds of things," she recalled. "And, of course, everything was taking longer than it was supposed to."[3] At 4:00 in the morning, they packed up and drove home, leaving the instruments on. Fifty minutes later, at 4:50, a gravitational wave arrived from 1.3 billion light years away, stretching and squeezing space itself, including the lasers, first at the Louisiana site and then seven milliseconds later at the Hanford facility. It was four days before LIGO's official reopening, and everyone involved with the project in the United States was sleeping. In Germany, it was 11:51 a.m., time for lunch. Marco Drago, a young physicist working at the Max Planck Institute for Gravitational Waves in Hannover, took a look at the data coming from LIGO. He immediately recognized the gravitational wave pattern predicted by the computer simulations. His excitement at being the first human to see evidence of gravitational waves was tempered by fears that this was a cruel artifact that looked similar to the expected signal or was a test released by the LIGO executives. Running a test during the warm-up period sounded like something the LIGO folks would do. One of Drago's colleagues contacted the LIGO operations room in Louisiana. David Reitze, the LIGO executive director, dropped his daughter off at school and went to his Caltech office, where he was greeted by a barrage of messages. "I don't remember exactly what I said," he told Nicola Twilley. "It was along these lines: 'Holy shit, what is this?'"[4] He knew he hadn't authorized a test; this could be it. The LIGO pioneers who had dreamed about this moment for 40 years were in their eighties now, and their dreams were about to come true. Rainer Weiss, age eighty-three, was on vacation in Maine. When he logged onto the computer and saw the gravitational wave signal, he yelled "My God!" so loudly his wife and grown son came running to see what had happened.

Everyone was sworn to secrecy while the LIGO researchers checked and rechecked. It wasn't a test, and it wasn't noise: it was the real thing, a gravitational wave. By February 2016 everyone involved was convinced, and a press conference was held to announce that Einstein's gravitational waves had been detected. At the same time, they published a paper in *Physical Review Letters* describing the findings. It had more than 1,000 authors.[5]

Some 1.3 billion years ago life on Earth was progressing from unicellular to multicellular organisms. At the same time, somewhere far away in our southern skies, two massive black holes were locked in a dance. One hole was 29 times heavier than the sun; the other was 36 times more massive than the sun, but with a diameter of less than 200 kilometers, it was unimaginably dense. They swirled around each other, faster and faster, in increasingly smaller orbits, reaching speeds close to the speed of light, until finally they crashed into each other. For a few moments, the result was the most radiant object in the universe. Space and time were distorted, and gravitational waves with energy equivalent to three solar masses were released. They sped away from the newly formed black hole, which was 63 times heavier than the sun and yet not much bigger than the state of Maine. They traveled in all directions, moving at the speed of light, compressing and stretching time and space as they went. While the gravitational waves were making their way through the universe, humans evolved from unicellular organisms to the stage where they could build a LIGO facility that would detect the passage of the waves on September 14, 2015.

How do we know all this? Well, besides his LIGO research, Kip Thorne is also known as the scientific adviser for the film *Interstellar*. In 2014 his book, *The Science of* Interstellar, was published, and in it he answers that question:

If we know the mass of a black hole and how fast it spins, then from Einstein's relativistic laws we can deduce all the hole's other properties: its size, the strength of its gravitational pull, how much its event horizon is stretched outward near the equator by centrifugal forces, the details of the gravitational lensing of objects behind it. Everything. This is amazing. So different from everyday experience. It is as though knowing my weight and how fast I can run, you could deduce everything about me: the color of my eyes, the length of my nose, my IQ.[6]

Within a year of the announcement of the detection of the gravitational waves, 84 scientific papers were published analyzing the data from the collision that must have occurred 1.3 billion years ago. According to Robert Ward of the Australian National University, in just a year, "We learned that gravity and light travel at the same speed, neutron star mergers are a source of short gamma-ray bursts, and that kilonovae—the explosion from a neutron star merger—are where our gold comes from."[7]

Since then the LIGO observatories have detected gravitational waves at least 40 more times. Surely many more waves have passed through us in that period, but they were too weak to be detected. Gravitational waves from the biggest and most distant black hole collision observed to date were recorded on July 29, 2017. The merger occurred five billion years ago, and in it nearly five times the mass of our sun was converted into gravitational waves.

Less than a month later, on August 14, 2017, a gravitational wave was detected at the two LIGO observatories in the United States as well as at the newly updated European Gravitational Observatory's Virgo detector, located outside Pisa, Italy. It was the first wave recorded at three different observatories, allowing the researchers to triangulate the location of the approximate birthplace of the gravitational waves.

On April 1, 2019, both LIGO and the Italian Virgo observatories restarted after undergoing a 19-month upgrade. The improvements have increased their sensitivity by at least 40 percent. Researchers have predicted that with the increased sensitivity, they will detect gravitational waves on a weekly basis. So far they have been proven correct. To celebrate and display their newfound confidence, the gravitational wave hunters have started an open public alert system. Reports of new gravitational wave observations no longer have to undergo lengthy validation delays, and the hope is that by alerting the astronomy community of new gravitational waves as soon as they are recorded, the astronomers can immediately use their electromagnetic telescopes to search for the source. In the first 5 months they reported the locations of 30 new events.

In 2019, the Indian government approved the construction of a new $277 million LIGO facility in Maharashtra state in western India. It will be the sixth LIGO observatory in the world. The new LIGO facilities let us examine large swaths of the universe that were invisible to us before. With six observatories, we will be able to scan larger sections of the sky and become more confident in detecting and triangulating gravitational waves.[8]

GRAVITATIONAL WAVES GET A NOBEL PRIZE

On October 3, 2017, just a little more than a year after the first gravitational wave was observed, the Nobel Prize in physics was awarded to Rainer Weiss, Barry C. Barish, and Kip S. Thorne "for decisive contributions to the LIGO detector and the observation of gravitational waves." Barry Barish is also from Caltech and joined the project fairly late, in 1994. He is responsible for many of the updates made to the LIGO system that made it sensitive enough to detect the gravitational waves. Professor Olga Botner, a member of the Nobel Committee for Physics, when interviewed by freelance journalist Joanna Rose about the 2017 Nobel Prize in physics, said that the laureates were not surprised when they got the phone call from the Nobel Committee. The discovery was so important that everyone was expecting them to get the prize, including the awardees themselves, but they were also humbled and gratified by the honor. All three stressed that this was a project that involved many scientists working together all over the world and that they wanted to acknowledge their collaborators. The project that had started in MIT and Caltech had grown to 90 universities across the world.[9]

Much as fluorescent proteins have provided biologists with important new information that demarcates the starting point for a swath of discoveries to come, so too has LIGO opened this door for astronomy. Previously, astronomers mainly relied on electromagnetic radiation (X-rays, visible light, microwaves, radio waves, etc.) to study the universe. Gravitational waves are completely unrelated to electromagnetic radiation; this was like comparing sight and sound. Szabolcs Marka, a Columbia University professor and LIGO scientist, says, "Everything else in astronomy is like the eye. Finally, astronomy grew ears. We never had ears before."[10] Of course, sound doesn't travel in a vacuum, so we can't hear the distant explosions; we have to rely on detecting the gravitational waves. Nevertheless, astronomers just got a new sense, the art of hearing some of the most energetic events in the universe: colliding black holes, exploding stars, merging neutron stars, and possibly even the big bang itself. These events are completely invisible to "traditional" observatories that are based on electromagnetic radiation. With LIGO/Virgo and their view of gravitational waves, a new window has opened up, allowing us to observe the universe in a way never possible before. "Until now, we scientists have only seen warped space-time when it's calm," Dr. Kip Thorne wrote *New York Times* journalist David Overbye in an email. "It's as though we had only seen the ocean's surface on a calm day but had never seen it roiled in a storm, with crashing waves."[11]

THE ROLE OF BIG SCIENCE

Big science is becoming more common due to increased specialization, the need for interdisciplinarity, and improved communication technologies. From 2014 to 2018, 1,315 papers with more than 1,000 authors were published, nearly double the hyperauthored papers published between 2009 and 2013. The LIGO project, with its mega-budget and thousands of researchers scattered all over the world, is an extreme example. The photographing of the black hole is another example of big science; it involved 200 people from 60 institutes and 20 countries and regions. These massive projects show how funding agencies, managers, and scientists from many different countries and cultures can come together to solve intricate problems.

Big science, however, is not able to solve every scientific puzzle we're up against, and some people are skeptical of its merits. James Evans and Lingfei Wu examined 65 million papers, patents, and software products released between 1954 and 2014.[12] According to Evans, "Big teams take the current frontier and exploit it. They wring the towel. They get that last ounce of possibility out of yesterday's ideas, faster than anyone else. But small teams fuel the future, generating ideas that, if they succeed, will be the source of big-team development."[13] The analysis showed that smaller groups are more likely to publish innovative research that may often take longer to be recognized. This was true for all decades studied and in all fields, from chemistry to social sciences. Evans thinks that one of the drawbacks of big science is that the large price tag and pressure to please funders drive researchers to safer areas of study. This is not necessarily a bad thing; we need to find the right mix of small and big science. Evans warns that we can't ignore small research groups, as they have a very important role to play. They provide the theories and breakthroughs that big science is built on. The next two examples of new breakthroughs in science (optogenetics and CRISPR) show how numerous small, disparate groups can combine ideas to make big breakthroughs.

Chapter Seven

Deep Learning

Deep learning is a subset of machine learning where artificial neural networks, algorithms inspired by the human brain, learn from large amounts of data.—Bernard Marr

While science is changing the way we will live, deep learning is changing the way we will do science. It is changing the way medicine is practiced, protein folding is simulated, and organic syntheses are designed.

In this book I have purposely deemphasized computer science, which needs another book in itself that I am not qualified to write. However, I include here a few pages about deep learning because it is going to change our lives and has already changed the way in which we do science. For more about deep learning, read Kai-Fu Lee's *AI Superpowers: China, Silicon Valley, and the New World Order*[1] and Eric Topol's *Deep Medicine: How Artificial Intelligence Can Make Healthcare Human Again.*[2]

Deep learning uses the often hidden information contained in vast data sets to solve questions of interest, particularly in games, speech and voice recognition, autonomous cars, science, and medicine. In *AI Superpowers,*[3] Kai-Fu Lee writes that artificial intelligence (AI), which encompasses deep learning, has surpassed the intellectual development stage, where its growth was dependent on a few extremely talented individuals making breakthroughs. Instead, it is now in the application phase. AI is a little like electricity after Thomas Edison and George Westinghouse brought electricity to our homes and inventors searched for new uses for the power source. Researchers in small start-ups and mega companies such as Baidu, Alphabet (Google's parent company), and Facebook are looking for new applications of deep learning and AI.

CHESS, GO, AND POKER: PROOF OF
CONCEPT APPLICATIONS OF DEEP LEARNING

In 1997, after years of programming, IBM's Deep Blue chess program beat the reigning world chess champion, Garry Kasparov, in six games. This was not surprising, because the application had been programmed using principles derived from strategies employed by grandmasters themselves. It learned endgames and openings. That knowledge, combined with the ability to evaluate millions of positions per second without getting tired or making a mistake, made it a fearsome but not invincible opponent. Chess programs and computers evolved throughout the following years. In 2016, Stockfish-8 was the world's computer chess champion. It evaluated 70 million chess positions per second and could draw on centuries of accumulated human chess strategies and decades of computer experience. It played efficiently and brutally, mercilessly beating all its human challengers without an ounce of finesse. Enter deep learning! On December 7, 2017, Google's deep learning chess program AlphaZero thrashed Stockfish-8. The programs played 100 games, with AlphaZero winning 28 and tying 72. It didn't lose a single game. AlphaZero only did 80,000 calculations per second, and it took just four hours to learn chess from scratch by playing against itself a few million times and optimizing its neural networks as it learned from its experience. AlphaZero didn't learn anything from humans or chess games played by humans. It taught itself, and in the process derived strategies never seen before. In a commentary in *Science* magazine, Kasparov wrote that by learning from playing itself, AlphaZero had developed strategies that "reflect the truth" of chess rather than reflecting "the priorities and prejudices" of the programmers. "It's the embodiment of the cliché, 'work smarter, not harder.'"[4]

Chess programs and chess machines have long been considered model systems for human reasoning. Alfred Binet, inventor of the first intelligence test in 1904, hoped that understanding how people played chess would unlock the secrets of human thinking. Alan Turing wondered if a chess engine might illustrate the differences between the potentialities of the machine and the mind.[5] However, it seems that playing chess is not the greatest model for human thought. AlphaZero can beat the world's best chess players because chess is a closed system with a fixed and fairly limited number of options.

Go soon replaced chess as a model of a system for human thought. Creating a competitive Go program was considered the new Mount Everest of computational challenges. The game Go was invented in China more than 2,500 years ago and is played by over 46 million people. In China, Go is much more than a game; it has a cultural and philosophical significance. It is still a closed system, but the number of potential positions is much greater than in chess. There are more than 10^{170} positions in Go, versus no more than

10^{50} in chess. Go has more potential moves than there are atoms in the universe. That didn't stop AlphaGo from teaching itself (with the help of 17 Go experts) how to play Go, and in March 2016 it beat South Korean Go legend Lee Sodel by four to one games. The games were watched by 280 million Chinese viewers and had a significant impact on the Chinese viewing audience and the government. According to Kai-Fu Lee, this was China's Sputnik moment. An American program beating a Go master got everyone's attention. By May 2017, when AlphaGo beat Chinese Go master Ke Jie in three 3-hour games, China had made research and development in AI a national priority. Just a year after AlphaGo beat Sodel, 48 percent of all AI global venture funding was from China, surpassing U.S. investments for the first time.[6]

Having overcome chess and Go, AI developers needed a new challenge. In chess and Go all players have the same information, while in poker players are dealt cards individually, there are multiple players with competing interests, and players can bid and bluff. The hidden information means that there is an information asymmetry, and beating professional poker players is therefore a different kind of challenge. It is more difficult and reflects real-life challenges. In July 2019 Pluribus, a program created by two Carnegie Mellon professors, played and beat the world's top 15 six-player, no-limit Texas hold 'em poker players in a 12-day session involving more than 10,000 hands. The program and results were reported in an article published in *Science*. Once trained, the program used just two CPUs. This was the first time a computer beat multiple elite poker players. As was the case with the chess and Go programs, Pluribus has the potential to teach human poker players something. In the *Science* paper the authors write, "Because Pluribus's strategy was determined entirely from self-play without any human data, it also provides an outside perspective on what optimal play should look like in multiplayer no-limit Texas hold 'em."[7] The program only looked ahead a few rounds, used a simplified bidding process, and turned out to be very good at bluffing. Chris Ferguson, a Pluribus victim and six-time World Series of Poker champion, said it was very difficult to pin the computer down on any kind of hand. It will be interesting to see what Pluribus does to online poker gambling. Noam Brown and Tuomas Sandholm, the authors of Pluribus, could clean up by letting the algorithm flex its muscles in the $3.6 billion online poker world.[8]

PROTEIN FOLDING: DEEP LEARNING TAKES THE NEXT STEP

I am a computational chemist. My students and I use programs developed and refined by many researchers over decades of work. These programs are very good at calculating structural changes that occur when we make small changes to known structures of molecules. However, a big challenge remains in computational chemistry: how to simulate protein folding. Misfolded proteins are most likely responsible for Alzheimer's disease, Parkinson's disease, cystic fibrosis, and Huntington's disease. According to Liam McGuffin, a computational chemist at Reading University in the United Kingdom, "The ability to predict the shape that any protein will fold in to is a big deal. It has major implications for solving many 21st-century problems, impacting on health, ecology, the environment and basically fixing anything that involves living systems."[9]

Every two years, the world's top computational chemists test the abilities of their programs to predict the folding of proteins and compete in the Critical Assessment of Structure Prediction (CASP) competition. In the competition, teams are given the linear sequence of amino acids for proteins for which the three-dimensional (3D) shape is known but hasn't been yet published; they then have to compute how those sequences would fold. In December 2018, 98 teams competed in the 13th CASP competition in Cancún, Mexico. The competition was won by the rookie AlphaFold team, the Google entry. It successfully predicted the structure of 25 out of 43 proteins, while the second-place system could only predict 3 protein structures. "For us, this is a really key moment," said Demis Hassabis, cofounder and CEO of DeepMind, the Google subsidiary responsible for creating AlphaFold. "This is a lighthouse project, our first major investment in terms of people and resources into a fundamental, very important, real-world scientific problem."[10] One of the reasons for AlphaFold's success is that it could use the Protein Database, which has over 150,000 experimentally determined 3D structures, to train itself to calculate the correctly folded structures of proteins. As with AlphaZero and AlphaGo, we don't exactly know what the program is doing and why it uses certain correlations, but we do know that it works. Besides helping us predict the structures of important proteins, understanding AlphaFold's "thinking" will also help us gain new insights into the mechanism of protein folding.

One of the most common fears expressed about AI is that it will lead to large-scale unemployment. AlphaFold still has a significant way to go before it can consistently and successfully predict protein folding. However, once it has matured and can simulate protein folding, computational chemists will be integrally involved in improving the program, trying to understand the underlying correlations used, and applying the program to solve important problems such as the protein misfolding associated with many neurodegenerative diseases. AlphaFold and its offspring will certainly change the way computational chemists work, but it won't make them redundant.

AI, MEDICINE, AND CHANGING EMPLOYMENT PATTERNS

Those working in other areas won't be as fortunate. In the past robots have been able to replace humans doing manual labor, and with AI our cognitive skills are also being challenged. According to an analysis by investment bank Morgan Stanley, autonomous trucks have the potential to save the industry $160 billion annually in labor, fuel, and productivity. They would also improve the safety and decrease the environmental impact of trucking. The report concludes: "Longhaul freight delivery is one of the most obvious and compelling areas for the application of autonomous and semi-autonomous driving technology" and "will be adopted far faster in the cargo markets than in passenger markets."[11] Once again China is leading in this area of AI applications. It has already started building highways designed for autonomous trucks.[12]

In recent years, at least 60 percent of my incoming introductory chemistry classes have been premed students. Many don't survive organic chemistry. In the United States, all premed students have to take organic chemistry, not because organic chemistry is useful in the practice of medicine, but because the skills needed to master organic chemistry are very similar to those used in medicine. Students in organic chemistry have to memorize many functional groups and reactions before combining them in new ways to make a previously unseen product. Similarly, medical diagnosis relies on gathering all the patient's unique symptoms and finding the underlying cause among the roughly 10,000 known diseases. I think you can see where I am going with this: medical doctors collect all the patient's blood work, symptoms, and other details; go to their vast, memorized database of known maladies to find the cause; and then proceed to treat the symptoms, a procedure perfectly suited to machine learning. I am sure that by the time the students in my current introductory chemistry class complete their residencies and start looking for jobs in the medical field, AI will have changed the way medicine is practiced. Deep learning algorithms will most likely be involved in many other aspects of modern medicine, such as the initial triage of patients and the analysis of diagnostic scans. The insured and relatively well-off patients will have small biosensors permanently embedded in their bodies, which will continuously monitor levels of key proteins, metabolites, minerals, and so forth and send this information to a central computer to process their vitals. Based on the data gathered from millions of patients, the AI system will probably know that we are sick even before we experience the symptoms ourselves; it will prescribe a treatment, monitor its efficacy, and adjust the dosage appropriately. If needed, the AI system will send the patient to the doctor. Health-care costs in the United States have been skyrocketing, yet the amount of time doctors spend with their patients has steadily decreased.

Perhaps the combination of medicine and deep learning, what Eric Topol calls deep medicine, will give doctors the time to care for their patients. [13]

Most researchers working at the interface of deep learning and medicine estimate that currently less than 5 percent of all the medical information available to doctors (big data) is being used. There are numerous reasons for this, foremost being that there is no common data format among the many different specialists, hospitals, and insurance companies; that we don't know what all the data mean; and that electronic health records were designed to facilitate billing, not to improve health. [14] The data are "dirty" and not as easy to interpret as in games and protein folding.

In the United States, researchers have developed algorithms that are at least as good as humans at interpreting X-rays, mammograms, and photos of skin cancers. However, it will take a while before deep learning will regularly be applied in medical diagnoses. If a radiologist makes a misdiagnosis, it affects one patient; however, if an AI program makes an error, it could impact hundreds of patients. This means there will be a high bar for the adoption of deep learning diagnostic programs. The medical field is averse to change, especially when it is not in medical professionals' best (personal) financial interests. The medical associations are protecting their members. The annual compensation of a radiologist in the United States is $400,000, which translates to a fair amount of political power. [15] Yann LeCun, the chief AI scientist at Facebook, disagrees with this approach; he thinks radiologists should embrace deep learning diagnosis because it will take care of all the boring, straightforward, simple cases and leave them with the interesting and complex ones. [16]

Most deep medicine will most likely come from China, which in turn will drive research in the United States. In China, Baidu has developed an AI-powered diagnostic program, RXThinking, to compensate for the fact that most well-trained doctors in China are concentrated in the wealthiest cities, resulting in large swaths of rural areas having few doctors. RXThinking acts as an automated triage station. Given a set of symptoms, the algorithm makes a series of diagnoses with a confidence rating for the diagnosis, which is then given to the doctor. The program uses over 400 million medical records and is continually updated with the latest medical publications. At this point (2019), the program is used to lighten the workload of doctors in rural areas and to provide the same high-quality diagnoses offered in expensive hospitals in the wealthy cities. [17]

AI diagnostics are making their way into the U.S. market, too. In late 2018, Amazon announced a collaboration with the Fred Hutchinson Cancer Research Center to evaluate "millions of clinical notes to extract and index medical conditions." Together they will use the data to train their machine-learning medical diagnostics algorithm, which already compares favorably with all other published diagnostic programs. [18] CVS has also partnered with an AI company to bring computerized diagnostics to its 1,100 minute clinics, at which the program will be used to diagnose and recommend the best care and treatment for their customers. From 2013 to 2018, AI health-care start-ups raised more money than any AI other sector.

At the beginning of this section I explain that premed students take organic chemistry because it trains them to think like medical doctors. If that is correct, then deep learning algorithms should also be able to master organic chemistry. Marwin Segler, an organic chemist at the University of Munster, Germany, trained them to do just that. He and his colleagues designed a deep learning program that used the more than 12 million known single-step organic reactions to design new small organic molecules.[19] Synthetic organic chemists think of themselves as artists, and I have often heard them, in classes and at conferences, describe organic syntheses using vocabulary borrowed from art critics. So it is particularly noteworthy that in double-blind tests, top organic chemists could not distinguish between published and AI-designed syntheses.

It may take a while, but I am sure deep learning will change the way science and medicine are done.

ARE WE ALL JUST DATA?

Because of AI's reliance on data, data have become a valued and desired asset. Everyone is collecting data. I particularly enjoyed the Fitbit announcement that they have 150 billion hours of heart-rate data. For tens of millions of people around the world, Fitbit can match their heart rate data with sex, age, weight, location, and sleep patterns. What a treasure trove for cardiologists, if they could get their hands on it!

Ten years ago, investors struggled to figure out Google, Facebook, and WeChat's business models. What were they making or selling? Today it's obvious: lots of data, which we are willing to give them for nothing, just for the privilege of using their products. Now we are stuck; if we don't want to give these companies our data, we have to give up using Facebook, Gmail, and Google.

Writing in *The Atlantic*, Yuval Harari warns us of the dangers of too much data being concentrated in a small number of AI companies: "Currently, humans risk becoming similar to domesticated animals. We have bred docile cows that produce enormous amounts of milk but are otherwise far inferior to their wild ancestors. They are less agile, less curious, and less resourceful. We are now creating tame humans who produce enormous amounts of data and function as efficient chips in a huge data-processing mechanism, but they hardly maximize their human potential. If we are not careful, we will end up with downgraded humans misusing upgraded computers to wreak havoc on themselves and on the world."[20] I think this is a rather extreme view, but not a totally unrealistic one.

Like science, or maybe because it is part of science, AI is progressing so quickly that it is very difficult to foresee all the consequences and to provide guidelines for its development. In an article in *Forbes* titled "Artificial Intelligence Regulation May Be Impossible," Michael Spencer argues that we might have to use AI to police itself. "Artificial Intelligence regulation may be impossible to achieve without better AI, ironically. As humans, we have to admit we no longer have the capability of regulating a world of machines, algorithms and advancements that might lead to surprising technologies with their own economic, social and humanitarian risks beyond the scope of international law, government oversight, corporate responsibility and consumer awareness."[21]

Much of deep learning relies on having lots of data to train the programs and to find hidden correlations. The big internet companies have lots of data, as do the U.S. and Chinese governments. Due to the number of people, connectedness, and lax privacy laws, the Chinese companies and government have a data edge. The *Economist* has even characterized China as the "Saudi Arabia of data." The reliance on data; well-paid programming jobs (starting salaries of between $300,000 and $1 million have led to jokes that the AI industry should institute a National Football League–like salary cap); and the ability to gobble up small, interesting start-ups have resulted in these big players having a significant lead in the AI arena. It is unlikely they will lose this lead, resulting in the gap between the haves and have-nots getting bigger and bigger. Perhaps this foreshadows the future of science.

Artificial intelligence has been in the news a lot recently because of its use in autonomous cars, facial recognition, and fake news. If it hasn't changed the way we live yet, it soon will. It has certainly changed the way science is done, as confirmed on the July 7, 2017, cover of *Science* magazine, which proclaimed: "AI Transforms Science." It has allowed scientists to create massive amounts of data and find their deeper meaning. For example, about 10^{60} different small molecules can be made in the lab. That is more molecules than there are atoms in the galaxy. Many of these molecules will have useful pharmaceutical properties. More than 60 start-ups and 16 pharmaceutical companies are using deep learning to scan this vast molecular space to discover new drugs.

Besides being an enormous aid to science, the growth of deep learning foreshadows many of the concerns and ethical dilemmas discussed in the next two chapters, on optogenetics and CRISPR.

Chapter Eight

Optogenetics

Using Light to Turn on Neurons One at a Time

At the Charles Stark Draper Laboratory in Cambridge, Massachusetts, biomedical engineer Jesse J. Wheeler has turned live dragonflies into drones. This is not a science fiction movie or an episode of *Black Mirror*. Each of Wheeler's dragonflies wears a tiny backpack that holds a microcomputer and flexible optic fibers. The tiny fibers run from the backpack into the dragonfly's nervous system.

Wheeler controls the dragonflies by activating and deactivating their individual neurons. Via the backpack, Wheeler uses the optic fibers to send flashes of light to specific neurons, turning them on or off. When the neurons fire, or turn on, signals travel through the dragonfly's nervous system. Wheeler has determined which neurons to fire to send signals that cause a dragonfly to change its direction of flight. Wheeler calls his remotely controlled insects DragonflEyes. "DragonflEye is a totally new kind of micro-aerial vehicle that is smaller, lighter, stealthier than anything that is manmade," he says.[1]

The process of switching neurons on and off using flashes of light is called optogenetics. As the name implies, it involves optics, the science of light, and genetics.

Scientists hope to do more than control the neurons of dragonflies and laboratory animals. They want to use optogenetic technology to activate and deactivate neurons inside the human brain. They don't want to create remotely controlled human zombies, but they do want to use optogenetics to better understand how the human brain works. The human brain is extremely complex. It has 85 billion neurons. (A dragonfly brain has fewer than 1 million neurons.) Neuroscientists say that optogenetics will help them map the complicated neural circuitry deep inside the brain.

WHAT IS OPTOGENETICS?

Optogenetics is a logical outgrowth of the fluorescent protein work described in chapter 5. Like fluorescent proteins, optogenetics is being used on a daily basis in labs all over the world. It is expanding our understanding of the brain (and more) and will be used in numerous medical treatments. I hope this chapter convinces you that this is a technique worth keeping an eye on, that optogenetics is one of the most interesting scientific breakthroughs of the decade.

Every day new research publications using optogenetics are published. Most applications are difficult to explain and not particularly sexy. However, there are many mind-blowing, controversial studies among them that could have been taken straight from the realm of science fiction. It's these types of applications that we hear about in the general media. Possibly the most striking of these reports were recent articles describing attempts to "create fearless soldiers." In January 2017, researchers at Yale University optically activated the predatory instinct in mice so that they hunted just about everything placed in their paths, including inanimate objects such as bottle caps.[2] In a similar vein, in July 2017, scientists at Zhejiang University in Hangzhou, China, used optogenetics to make timid mice aggressive.[3] How did they do this?

Prior to optogenetics, the closest neuroscientists could come to this goal was to stimulate brain cells (neurons) with electrodes. But even the finest electrodes trigger thousands of neurons at once, never individual neurons. In 1999, Francis Crick, of DNA fame, was the first to suggest, in a paper titled "The Impact of Molecular Biology on Neuroscience," that light could be used to excite individual neurons, but he was careful to qualify his suggestion, saying that it was a "far-fetched" idea.[4]

That changed in 2005, when Karl Deisseroth pioneered the science of optogenetics. With optogenetics, scientists use light and proteins found in algae and bacteria to turn individual neurons on and off instantly. This is how it works. Certain kinds of algae have whiplike tails, called flagella, which they use to swim toward and away from the sunlight. These algae are called flagellates, and flagellates of the genus *Chlamydomonas* have been key to the science of optogenetics.

If we place a *Chlamydomonas* alga in a large aquarium in a darkened room, it will swim around aimlessly. But if we turn on a lamp, the alga will swim toward the light. The single-celled flagellates don't have eyes. Instead, they have a structure called an eyespot that distinguishes between light and darkness. The eyespot is studded with light-sensitive proteins called channel-rhodopsins. These proteins are particularly sensitive to blue light. Blue light travels farther through water than any other type of light. It is the most common light in aquatic environments, and most marine organisms see/use blue light.

Each channelrhodopsin contains an interior channel. When blue light shines on a channelrhodopsin, the channel opens and sends calcium ions into the eyespot, stimulating the flagella. The movement of calcium ions into the eyespot is very similar to the way calcium ions flood into neurons when they fire: the concentration of the calcium ions changes at least a hundredfold when neurons fire.

In 2003, Peter Hegemann and Georg Nagel, biophysicists at the Max Planck Institute for Biochemistry in Frankfurt, Germany, studied the light-sensitive behavior of *Chlamydomonas*. They discovered the channelrhodop-sin gene, isolated it, and incorporated the DNA into frog egg cells and human kidney cells in petri dishes. These genetically engineered cells now held the instructions for making the channelrhodopsin proteins themselves. When the researchers shone a blue light on the cells, the channelrhodopsins opened up, and calcium ions flooded the cells.[5]

Since calcium ions increase when neurons fire, the next step was to insert the gene for making channelrhodopsins into the neurons of laboratory ani-mals. Karl Deisseroth and his team at Stanford University first did this in 2005. They inserted the gene for making channelrhodopsin into rat neurons and studied the neurons in petri dishes. When the scientists shone a pinpoint beam of blue light on these neurons, the channelrhodopsins opened up, cal-cium ions flooded through the neurons, and the neurons fired.[6] The science of optogenetics was born. With optogenetics, scientists can stimulate specific small groups of neurons selectively and repeatedly. Joshua Sanes of Harvard University has argued, "Optogenetics has more than anything else let people play the piano in the brain, as opposed to just slamming their whole forearm down on all the keys."[7]

Once they had found a protein that could make neurons fire, neuroscien-tists also wanted to find a protein that would make neurons stop firing. They found one in a type of bacteria that thrives in salt flats. The protein is called halorhodopsin. Like channelrhodopsins, halorhodopsins contain interior channels. When yellow light shines on the halorhodopsins in bacteria, the channels open, and chloride ions flow through the bacteria. Chloride ions are negatively charged ions. They neutralize the activity of positively charged ions, such as calcium ions. Imagine a neuron that's firing; it is filled with calcium ions. But if chloride ions flood the firing neuron, these ions will negate the action of the calcium ions, and the neuron will stop firing.

SWIMMING WORMS AND CIRCLING MICE: PROOF OF CONCEPT OPTOGENETICS EXPERIMENTS

Caenorhabditis elegans, a tiny roundworm, has only 302 neurons in its nervous system, so the system is easy to study. Scientists have studied these neurons, and they know what each one of them is responsible for. To prove that halorhodopsins switch off neural activation, Karl Deisseroth and his team genetically modified the neurons responsible for movement in some of these worms. They genetically engineered these neurons to produce halorhodopsins.

C. elegans normally lives in soil, and when placed in water it has to swim rhythmically and constantly to avoid drowning. The only way to stop the worm from swimming is to switch off the neurons responsible for that movement. The researchers placed a genetically modified *C. elegans* in a water-filled petri dish. With no light shining on it, the worm swam as it normally does. When the researchers turned on a yellow light, the channels in the halorhodopsins in the worm's neurons opened up, chloride ions flooded into the neurons that control movement, and the neurons stopped firing. When the neurons stopped firing, the worm stopped swimming. This was not in the worm's best interest; it started to drown. When the yellow light was switched off, the worm started swimming again.[8]

The nervous system of *C. elegans* has few similarities to the human nervous system, so turning the roundworm's neurons on and off doesn't tell us much about switching human neurons on and off. Mouse brains bear a closer resemblance to our own, so controlling neurons in mouse brains provides insight into the human brain.

Karl Deisseroth and his team inserted the gene for creating channelrhodopsin into neurons responsible for movement in mice. Rather than injecting the gene into mouse embryos, the scientists put the gene into a benign virus and injected the virus into the brains of laboratory mice. To direct light to the mouse brains, they implanted fiber optic cables (thin glass light-carrying fibers) into the brains of the genetically modified mice. Then the scientists sent blue light through the fibers and into the animals' brains. When the light hit the channelrhodopsins in the genetically modified neurons, calcium ions flooded the neurons and they fired.

Feng Zhang was a graduate student responsible for many of the early optogenetics breakthroughs in Deisseroth's lab. He is now a professor at MIT and has been a central player in the development of CRISPR. In a 2007 demonstration of optogenetics, Zhang shone light on genetically modified motor neurons on the right side of a mouse's brain. These are the neurons that control movement on the left side of the mouse's body. When he turned on the light, the mouse started running in circles to its left. When he shut off the light, the mouse stopped running.[9]

OPTOGENETICS AND MEDICINE

According to Ed Boyden of MIT, "The eye, which can access light from the outside world, is a perfect test bed for the use of optogenetic tools for treating [human vision] disorders."[10] In April 2011, a team led by Boyden and neuroscientist Alan Horsager of the Keck School of Medicine at the University of Southern California carried out optogenetics studies on blind mice.

The mice had retinitis pigmentosa, a disease in which light-sensitive cells in the retina are destroyed and the brain no longer receives visual information from the eyes. Horsager and his lab partners used a virus to insert the gene for producing channelrhodopsins into the damaged cells on the surface of the mice's retinas. The scientists hoped that the light-sensitive channelrhodopsins would be produced in the eye and take over for the destroyed retinal cells, thereby enabling the mice to sense light.

To test whether the experiment had worked, the scientists placed mice from three different groups in a water-filled maze with an illuminated exit. Mice with untreated retinitis pigmentosa were blind, so they swam around randomly and eventually happened upon the exit. Sighted mice and previously blind mice that had been genetically modified to produce channelrhodopsins saw the light and headed straight for the exit. The test showed that the channelrhodopsins enabled the previously blind mice to sense light.[11]

Such research has important implications for humans. More than 100,000 Americans have lost their sight to retinitis pigmentosa. Doctors want to use channelrhodopsins to enable them to sense light, just as the blind mice did. In 2016, David Birch of the Retina Foundation of the Southwest in Dallas, Texas, started testing the procedure on humans. He used a benign virus to insert the gene for making channelrhodopsins into subjects' eyes. The goal was not to restore perfect, multicolor vision to those living with retinitis pigmentosa. Instead, Birch hoped to give patients some improvement in vision, such as "being able to [safely] cross the road."[12] This was the first time optogenetics had been used to treat humans.

In addition to the Retina Foundation of the Southwest, several U.S. and French companies hope to use optogenetics to cure retinitis pigmentosa. Channelrhodopsin responds to blue light. Yet the light we see all around us covers the entire visible spectrum, from violet on one end to red on the other. So in addition to giving the channelrhodopsin gene to patients with retinitis pigmentosa, optogeneticists want to develop glasses that will enable patients to respond to light of other wavelengths as well.

To use optogenetics to treat diseases, one needs to precisely focus light on specific neurons. In treating retinitis pigmentosa in humans, getting light into the retina is not a problem because light naturally enters the eye, enabling us to see. But most other optogenetic applications require a small, localized light source inside the organ of interest, which is normally the brain. Biomedical engineers are already developing such lights. They will be implanted into patients' brains, remotely controlled, and turned on and off as needed to fire specific neurons and cause others to stop firing. Once these devices are perfected, they will allow neuroscientists to test more optogenetic treatments on humans.

In addition to receiving implanted lights, humans will need to undergo genetic modification to receive optogenetic treatments. The most common technique would likely be to add the channelrhodopsin gene into a harmless virus that would carry the gene to specific neurons. Some researchers are nervous about this step. They fear it might not work at all or could have unintended side effects. "It is important to remember that animals are not people, and so there will always be some risks associated with the introduction of nonhuman DNA into humans that cannot be predicted in the lab," says Stanford's Lauren Milner.[13]

Nevertheless, it's clear that optogenetics will play a role in medicine in the future. Doctors frequently treat brain disorders such as depression using pharmaceutical drugs. But the drugs don't target neurons precisely. David Anderson, a biology professor at the California Institute of Technology, compares these drugs to a sloppy oil change. "If you dump a gallon of oil over your car's engine, some of it will dribble into the right place, but a lot of it will end up doing more harm than good," he says.[14] With optogenetics, however, neuroscientists are gaining a more precise understanding of which neurons to target to treat specific disorders and diseases. This knowledge will help researchers create medicines that can treat patients more effectively.

Most exciting of all, optogenetics might hold the key to treating debilitating and deadly brain diseases, such as Alzheimer's and Parkinson's. Some say the technology is poised to completely transform all of medicine. That is because optogenetics isn't just used in the brain; for example, in the few weeks that I was writing this chapter, researchers have used optogenetic techniques to control the heartbeat of mice[15] and regularize the bowel movements of constipated mice.[16] However, we should remember that it is a difficult and long journey from proof of concept experiments in mice and other model organisms to actual applications in doctor's offices and hospital operating rooms. Furthermore, for every application we read or hear about in the press, there are hundreds of equally important, less spectacular, less sexy applications of optogenetics, in which light and genetics are used as tools to find new puzzle pieces in research projects in labs all over the world.

I started this chapter with remotely controlled dragonflies acting as live microdrones and end it with red-light-induced fruit-fly ejaculations. Yes, I admit to choosing unusual experiments to illustrate my points, and I am aware that the use of these examples may skew your perception of science, which the vast majority of the time does not rely on fly ejaculation, but I hope that these examples illustrate the diversity of experiments that can be performed with optogenetics. Think of optogenetics as a tool like a microscope. Microscopes are continually being improved and used all the time, but most of the images are only interesting to the researchers themselves, and only a small percentage of the images make it into the mainstream media. The situation is the same for optogenetics.

It is true that we, myself included, are attracted to stories that are off the wall or personally relevant to us. As discussed in later chapters, this can lead to fake news spreading faster than real news and to our perceptions of science being skewed by very applied, outlandish, or speculative experiments. For example, we don't read about the thousands of experiments that have been conducted to devise new and improved channelrhodopsins, but occasionally we will read about some of their applications.

Remember that channelrhodopsins were isolated from motile algae that live underwater, and because blue light travels the farthest in water, these proteins are reactive to blue light only. Now take your hand and hold it over a flashlight. The only color you will see coming through your hand is red light. All the other colors are absorbed. Unlike in water, it is red light, not blue light, that penetrates your flesh the most. That is why channelrhodopsins sensitive to red light have been created. Many experiments using red-shifted channelrhodopsins have been conducted, including the red-light-induced fruit-fly ejaculations previously alluded to.

How and why would researchers be interested in male fruit-fly orgasms? Reproduction is crucial to the survival of all species. Sex is energy depleting and tiring. So why are animals interested in reproduction? Well, the answer seems obvious: they enjoy sex. However, it is commonly said that mammals are the only organisms that have intercourse for fun. So why do fruit flies have sex? To ensure that animals mate despite the energy costs, they must have a reward system that makes mating worthwhile. Interestingly, when they do not get their mating rewards many animals, including fruit flies, will get their rewards in other ways, such as imbibing some alcohol. Galit Shohat-Ophir and Shir Zer-Krispil at Bar-Ilan University in Israel were interested in finding out if ejaculation was a large part of the male fruit fly's reward system and whether natural reward systems could override alcohol addiction. "We wanted to know which part of the mating process entails the rewarding value for flies," said Shohat-Ophir, who was responsible for the project. "The actions that males perform during courtship? A female's pheromones? The last step of mating which is sperm and seminal fluid release?"[17] They created 12 genetically modified fruit flies that had the red-shifted channelrhodopsin in the neurons responsible for ejaculations in male fruit flies.[18] The researchers used the red-shifted version of channelrhodopsin because fruit flies can't see red light and wouldn't be influenced by seeing the red light, so any behavioral changes would clearly be a result of the presence of the channelrhodopsin. Normally fruit flies only ejaculate after a complex courtship ritual that involves stroking and mating with a female fruit fly. Thanks to optoge-netics, the Bar-Ilan fruit flies ejaculated after 30 seconds of red light expo-sure. They seemed to enjoy it, because given the option, they stayed in the "red light district" for up to three minutes, where they averaged seven or-gasms per minute. Orgasm-deprived males were attracted to ethanol laced fruit, while those that had experienced the joys of the orgasm-inducing red lights were impervious to the alcohol reward system. It seems that these tiny, annoying fruit flies have a lot in common with us "sex for fun" humans. Shir Zer-Krispil, the graduate student who did most of the work, says, "I definite-ly think animals have pleasure. It's hard if you define pleasure from a human point of view, but it [comes down to] very basic machinery that even simpler animals have."[19]

Here are some of the newspaper and website headlines related to this research: "Researchers Discover Fruit Flies Love Having Orgasms";[20] "Male Fruit Flies Love to Cum, and Turn to Alcohol If They Can't";[21] "Fruit Flies Have Orgasms—and, Apparently, They're Amazing";[22] and "Fruit Flies Have Mind-blowing Orgasms, Israeli Scientists Prove—And If They Haven't Gotten Laid, Fruit Flies Demonstrate a Desire to Get Drunk."[23]

As breakthroughs go, I think optogenetics has reached its teenage years. It has become a commonly used tool, some start-up companies have been formed (e.g., Circuit Therapeutics, RetroSense Therapeutics, Optologix), and numerous awards have been given to its discoverers, but in order for us to understand its full potential, it still has to mature.

Boyden and Deisseroth, two of the inventors of optogenetics, are excited about where the technology might lead. Boyden says, "In the future optoge-netics will allow us to decipher both how various brain cells elicit feelings, thoughts, and movements—as well as how they can go awry to produce various psychiatric disorders."[24] Deisseroth adds, "As a psychiatrist I'm hop-ing that we can continue understanding these deep questions about anxiety [and] depression . . . and get to a level where we can get to the nature of a patient's problem very precisely. . . . We want to be able to . . . pinpoint exactly the biology behind what patients are intensely suffering from."[25]

Chapter Nine

CRISPR
(Clustered Regularly Interspaced Short Palindromic Repeats)

At this point the only thing CRISPR won't be able to do is open your refrigerator door!—John Oliver

In February 2016, U.S. director of national intelligence James Clapper presented the annual worldwide threat assessment. In it he listed six weapons of mass destruction (WMDs) that were considered a global threat by the U.S. intelligence community. North Korea developing WMDs, China "modernizing its nuclear forces," and chemical weapons in Syria and Iraq were expected threats. A new, unexpected WMD that he listed was genome editing.[1] Seeing their work listed as a WMD was a big surprise to the scientific community and the researchers involved in developing CRISPR. It is CRISPR that is responsible for "genome editing" developing from an expensive, tedious, and fairly esoteric technique to a threat that has come to the attention of the director of national intelligence.

In sharp contrast, on May 22, 2018, in his remarks to the Alliance for Regenerative Medicine's annual board meeting, the commissioner of the FDA, Scott Gottlieb, said, "We're at a key point when it comes to cell and gene therapy. These therapies have the potential to address hundreds, if not thousands, of different rare and common diseases. For a long time, they were largely theoretical constructs. Now they're a therapeutic reality. And it's my expectation that they will soon become the mainstay of how we treat a wide range of illness."[2] This is the type of statement that the scientific community was expecting, describing CRISPR as a technique used to cure diseases, not as a bioterror weapon.

What is this technique with the strange, catchy, slightly cute name? Why does it inspire nightmares in the director of national intelligence and hope in doctors treating a myriad of genetic diseases? How does it reflect the state of science?

CRISPR-Cas9, commonly known as CRISPR, is a gene editing tool. It is the Microsoft Word for the book of life.

The molecular machinery in the cells of most living organisms makes proteins using instructions encoded in the DNA of the cell. DNA is like a scientific cookbook that contains all the recipes required to make every kind of protein found in an organism's body. Scientists call these recipes "genes." Every cell has all the recipes needed to make every protein. When a cell in your finger needs to make more muscle, the molecular machinery of that cell will find the recipes for the required proteins and make them. The complete set of instructions on how to make all the proteins in the body is called the "genome" (in essence the cookbook). The human genome has roughly 20,000 genes. Humans are made up of about 10 million million (10×10^{12}) cells. Each cell has a nucleus, and in each nucleus is a complete set of instructions.

At birth every cell has a complete recipe book; it will never get a new one, so the book is kept in a very safe place, in the nucleus of each cell. Red blood cells are the only exception; they have no nucleus. The mitochondria, the energy-producing organelles of cells, contain their own mini-recipes (just 16,000 nucleotides), which are inherited from the mother exclusively. When a new protein needs to be made, the appropriate gene is found and the instructions are copied. Once the complete gene has been read, a copy of the protein recipe is sent as messenger RNA to the ribosome, where the protein is expressed or made. Making copies of the gene in the nucleus and then sending them to the ribosome might seem rather laborious, but it protects the recipe book from the harsh chemical environment of the cell. The process is rigidly controlled. For example, although the genes with the instructions for producing proteins required in the eye are found in every cell in the body, they are copied and translated only in the eye, and only when they are needed. If there is a breakdown in this process or the recipes are corrupted, sickness and disease follow.

CRISPR is the molecular word processor that allows us to find not just a gene, but a very specific part of a gene, change it, delete it, make the cell express more or less of its gene product, or even add a completely foreign gene. With CRISPR scientists can change, remove, and add recipes, and they can even modulate how often the recipes are used. This powerful technique can modify every one of those 10 million million cells, and it has what seems to be limitless power.

Like the fluorescent proteins discussed in chapter 1, the utilization of the CRISPR-Cas system as a gene editing system is based on basic research.

Without the basic research there would be no CRISPR-Cas gene editing system.

CRISPR HISTORY

The CRISPR story starts in 1987, when Yoshizumi Ishino and his coworkers discovered a strange stretch of DNA in *E. coli*, a commonly studied gut bacterium. The DNA segment consisted of two alternating sequences: a palindromic, 29-nucleic-base-pair section that was always the same, called the repeat, followed by a distinctive sequence of 32 unique base pairs, called the spacer sequence. (As an aside, my favorite palindromes are "Sex at no/on taxes"; "Do geese see God?"; "Are we not drawn onward we few, drawn onward to new era"; "Borrow or rob"; "Mr. Owl Ate My Metal Worm"; and "Satan, oscillate my metallic sonatas.") Ishino published the sequence but admitted that the purpose of the 29-base palindromic repeat and the 32-base-pair unique spacer repeating sequence was unknown. At the time Ishino published his work, DNA sequencing was expensive and time consuming.

By 2002, the sequencing methods were cheaper and more common, and Ishino's repeat sequences were found in nearly half of all bacteria and most of the archaea that had been sequenced. At this point there were enough puzzle pieces for Ruud Jansen at Utrecht University to come up with a great acronym, CRISPR (Clustered Regularly Interspaced Short Palindromic Repeats) and later to realize that there was another set of genes close to, and associated with, the CRISPR sequence, the CRISPR associated gene (*Cas*). The proteins encoded by these genes, the Cas proteins, cut DNA. The missing piece to understanding the purpose of CRISPR was the discovery that the unique 32-nucleic-base spacer sequences were all fragments of viral genes that had been incorporated into the bacteria's DNA.

Viruses known as bacteriophages have been attacking bacteria for billions of years. They are by far the most common type of organism on Earth. A single teaspoon of seawater contains more bacteriophages than there are people in New York City. On Earth they cause a trillion trillion infections a second. Forty percent of marine bacteria are killed by bacteriophages.

All the pieces that were needed to solve the CRISPR puzzle were on the table: CRISPR was composed of palindromic repeats of DNA that were interspersed between 32-base-pair fragments taken from viruses, each fragment came from a different virus, and each CRISPR sequence was associated with a gene-cutting Cas protein. Eugene Koonin put all the pieces together while he was at the National Center for Biotechnology Information in Bethesda, Maryland, and came up with a bacterial defense system composed of an album of mug shots (CRISPR) and a hunt-and-destroy protein (Cas).

Koonin proposed that CRISPR is a bacterial defense system composed of two parts. The first part is a stretch of DNA that acts as an album of vanquished foes. When the bacterium overcomes an invading virus or another bacterium, it snips out a section of the defeated invader's genetic material and places it in the album. These genetic mug shots are separated by palindromic stretches of DNA, generating a stretch of DNA with CRISPR. The second component of the bacterial defense system is a search-and-destroy weapon. Each genetic mug shot has a search-and-destroy protein, also known as a CRISPR associated (Cas) protein, associated with it. These Cas proteins circulate in the cell, and when they encounter a stretch of genetic material corresponding to their genetic mug shot, they kill the invader by cutting its genetic material (DNA and RNA).

Much of modern science is derived from natural processes, so it should come as no surprise that the molecular weapons used in these massive unseen wars between viruses and bacteria can be repurposed for our use.

It took research groups spread all over the world 20 years to get to that point. All the research was basic research, done to understand why bacteria had this strange repeating DNA motif. After 20 years, some practical applications of the CRISPR-Cas system were starting to become apparent.

In 2007, Koonin's hypothesis that CRISPR is a bacterial defense system was confirmed by Danisco, a Danish food and beverage company. At first this may seem to be an odd participant in the CRISPR research puzzle, but a few facts about the milk-fermenting bacterium *Streptococcus thermophiles* should make clear why Danisco was interested in investigating bacterial defense systems. *S. thermophiles* is used by Danisco and other dairy producers to make yogurt and mozzarella and other cheeses. According to the U.S. Department of Agriculture (USDA), 1.02 billion kilograms of mozzarella cheese and 621 million kilograms of yogurt were produced from *S. thermophiles* in 1998. The annual market value of this bacterium is over $40 billion. Most dairy products contain some *S. thermophiles*, and humans ingest over a billion trillion live *S. thermophiles* a year.[3] This milk-fermenting bacterium that is critical to global dairy production is vulnerable to viral attacks. Viral infections of *S. thermophiles* are the largest cause of incomplete fermentation and lost production in the dairy industry.

French and American researchers at Danisco were interested in understanding CRISPR and determining if they could use the bacterial defense system to make *S. thermophiles* resilient against viral attacks. In their experiments, the research groups of Rodolphe Barrangou and Philippe Horvath purposefully infected a culture of *S. thermophiles* with two different strains of viruses taken from yogurt samples. Most of the *S. thermophiles* died, but those that survived were resistant to reinfection by the same viral species. Sequencing of the resistant bacteria revealed CRISPR sequences with a new fragment of viral genetic material inserted between two palindromic spacer fragments. The surviving bacteria had taken a genetic mug shot of the vanquished virus and placed it in their own DNA so that they and their offspring would be resistant to that strain of virus. Removing the viral genetic material from the CRISPR sequence resulted in bacteria that were once again vulnerable to attacks by the virus associated with that sequence. The genetic mug shot was crucial to *S. thermophiles* resistance. Today many manufacturers use *S. thermophiles* cultures that have CRISPR sequences that will protect them from the most common viral outbreaks. According to Barrangou, who did this research at Danisco USA, "If you've eaten yogurt or cheese, chances are you've eaten CRISPR-ized [yep, it is a verb now] cells."[4]

Jennifer Doudna, a biochemist with extensive experience working with RNA at the University of California, Berkeley, started working with CRISPR in 2006. She was attracted to the field because the CRISPR system uses RNA to find the viral mug shot. At a 2011 American Society for Microbiology meeting in San Juan, Puerto Rico, she met Emmanuelle Charpentier for the first time. Charpentier was an associate professor at the Laboratory for Molecular Infection Medicine Sweden at Umeå, where she worked on a CRISPR associated protein called Cas9. Doudna and Charpentier had complementary skills. While they were walking around the old town of San Juan, Charpentier convinced Doudna that Cas9 was responsible for finding the DNA sequence corresponding to the mug shot and cutting it. Doudna was intrigued and agreed to take a closer look at the role Cas9 played. Charpentier worked with Cas9 in *Streptococcus pyogenes*, the bacteria that cause strep throat and flesh-eating disease. Rather than send Doudna these dangerous bacteria, she overnighted her the DNA encoding the CRISPR-Cas9. Back in Berkeley, Doudna got her students to express the *Streptococcus pyogenes* genes in *E. coli* so that they could take a closer look at the CRISPR-Cas9 system. Up to this point Doudna and Charpentier were only interested in the mechanism of the CRISPR-Cas9 bacterial defense system. They were not thinking about applications. But the more Doudna learned, the more obvious it became that this bacterial system could be co-opted to edit DNA. Cas9 always searched for, found, and cut the DNA that matched the genetic mug shot in CRISPR. If her group could replace the spacers (genetic mug shots) in CRISPR with a DNA sequence of their own design (a fake mug shot), then the CRISPR-Cas9 system would search for this sequence and cleave that portion of the DNA. They were successful, and with some tweaking CRISPR-Cas9 was converted to a gene-editing technique. The first gene they tried to edit was the GFP gene. In their proof of concept experiment, CRISPR-Cas9 cut the GFP gene at the exact location defined by Doudna. She was delighted; CRISPR-Cas9 "was the perfect bacterial weapon: a virus-seeking missile that could strike quickly and with incredible precision."[5]

Doudna and Charpentier wrote up their results in a paper titled "A Programmable Dual-RNA–Guided DNA Endonuclease in Adaptive Bacterial

Immunity." Although the title only addresses the bacterial defense system and is probably incomprehensible to most nonscientists, the introduction to the work clearly states the importance and utility of the CRISPR-Cas9 system as a gene-editing technique: "Our study further demonstrates that the Cas9 endonuclease family can be programmed with single RNA molecules to cleave specific DNA sites, thereby raising the exciting possibility of developing a simple and versatile RNA-directed system to generate dsDNA breaks for genome targeting and editing."[6] Basically, this could be a way of editing genes that was much easier than anyone had previously imagined. The authors submitted their manuscript to *Science*, whose editors saw the potential of CRISPR and fast-tracked the publication. It came out 20 days after being submitted, and Doudna's life was changed forever. At more or less the same time, she filed a patent application for the CRISPR-Cas gene-editing system.

Let's go back to the puzzle analogy I presented in chapter 4 and apply it to the CRISPR-Cas system. In the late 1980s, when Ishino described CRISPR's odd repetitive DNA sequence, he laid out the first CRISPR puzzle piece. By the time Jansen named the sequence CRISPR in 2002, numerous pieces had been placed in the puzzle, but it was still fairly esoteric. In 2007, when Koonin and some research groups worked out that CRISPR was part of a bacterial defense system, things got more interesting, and more researchers started looking at all the puzzle pieces. When Doudna and Charpentier started collaborating on Cas9's function, researchers all over the world were playing with the CRISPR puzzle. Researchers with backgrounds in RNA, DNA, DNA/RNA cutting (endonucleases), fermentation chemistry, and infectious diseases were trying to find pieces to the CRISPR puzzle. So it should come as no surprise that at the same time Doudna and Charpentier were focusing their research on Cas9's role in DNA cleavage, another group was also investigating Cas9 and was aware of the potential of CRISPR-Cas9.

Virginijus Šikšnys is a molecular biologist at Vilnius University in Lithuania with a research background in restriction endonucleases (proteins that cut DNA). When he heard about the CRISPR-Cas system's role in bacterial defense, he started studying the CRISPR associated proteins (Cas) that do the DNA cutting. In 2012, he and his students determined that Cas9 was the protein that was doing the cutting. Šikšnys wrote up his results and submitted them to the journal *Cell* in April 2012. The editor rejected the manuscript without sending it out for review. Šikšnys, confident in his work and its importance, submitted the manuscript to the *Proceedings of the National Academy of Sciences (PNAS)*. The paper, "Cas9–crRNA Ribonucleoprotein Complex Mediates Specific DNA Cleavage for Adaptive Immunity in Bacteria," was submitted before the Doudna and Charpentier *Science* paper was published but needed some revisions and was only published three months after the Doudna paper appeared. Like Doudna, Šikšnys also foresaw the potential of the CRISPR system. In the *PNAS* abstract, he writes, "These findings pave the way for engineering of universal programmable RNA-guided DNA endonucleases."[7] But because it was published after the *Science* paper, his article received less attention and fewer citations. In an interview with *Wired* magazine, Šikšnys says, "Yes, I think of course my lab deserves credit because what we discovered was done independently in two labs," but he adds, "It's a very competitive field. It's part of the game."[8] Had Šikšnys been at Berkeley, would his initial submission to *Cell* have received the attention it deserved?

Chapter 7 introduced Feng Zhang, the graduate student working in Karl Deisseroth's optogenetics laboratory at Stanford. He used light to make rats run around in a circle, a memorable demonstration of optogenetics' power. When the Doudna and Charpentier *Science* paper was published, he was working as a professor of neuroscience at MIT and also had an appointment at the Broad Institute. Like Doudna, Charpentier, and Šikšnys, he was using the CRISPR-Cas9 system to edit DNA, but while the others did all their editing in solution, Zhang was slicing and dicing DNA with CRISPR-Cas9 in human cells. In January 2013, Zhang published his own *Science* paper, which described how CRISPR-Cas9 could be used for gene editing human and mouse cells.[9] A little later in January, Doudna also published a paper describing her new work with CRISPR-Cas9 in human cells.[10]

The human genome has 3.2 billion nucleotides. CRISPR-Cas9 will search through the genome, find the ~20 nucleotide target sequence, and cut it at the precise desired point. In December 2012, although Jennifer Doudna had applied for a patent seven months earlier, Feng Zhang asked his employers, MIT and the Broad Institute, to file a patent on his behalf because he considered his gene-editing tool to be superior to and significantly different from Doudna's. The Broad Institute lawyers, knowing that Doudna's claim was pending, paid an additional fee to accelerate their patent application and were granted a CRISPR-Cas9 patent before Doudna was awarded hers. The result has been a closely watched legal battle between the Broad Institute and the University of California, Berkeley. In round one, the court awarded the Broad Institute the patent based on the first to the line principle, but Berkeley has appealed. On February 8, 2019, the patent office announced that it would forthwith accept the University of California's original patent for CRISPR in all kinds of cells. This has apparently muddied the waters even more, and no one is quite sure who will get paid the patent rights when human therapies based on CRISPR-Cas9 reach the market.[11] The stakes aren't as high as they were first thought to be; most initial estimates were in the billion dollar range, but they have now dropped to annual revenues in the $10 million range due to the advent of new systems that don't use Cas9 and would not be covered by either patent. Still, that is no small chunk of change to give up.

All the major players have started their own CRISPR-based companies. Doudna founded Caribou and Intellia, Charpentier CRISPR Therapeutics, and Zhang Editas Medicine, and Šikšnys has a license agreement with Du-Pont.

Science in today's world is a complicated affair. Many researchers from all over the world in many disciplines are collaborating (largely virtually, by email and video calls), competing, starting their own companies, and yet always trying to understand more about nature. Often the research goes nowhere, often it goes in unexpected directions, and sometimes it leads to the most excitingly splendid conclusions. CRISPR-Cas9 is one of these cases; it started with a weirdly repeating genetic palindrome, matured via mozzarella and yogurt, and finally blossomed into a contested gene-editing tool.

All these companies and patent disputes might give the impression that research today is a cutthroat, winner-take-all endeavor. But like most things, the situation is not so simple, and when it comes down to it, most scientists are interested in the advancement of science above all else. Both Doudna and Zhang have deposited their CRISPR plasmids (plasmids are fragments of double-stranded DNA that typically carry genes and can replicate independently from chromosomal DNA) in the Addgene repository, so that they can be shared with researchers all over the world. Addgene's open source philosophy is an interesting reflection of science today. It is a nonprofit that was founded by Melina Fan, her brother, and her husband in 2004. As a graduate student, Fan was frustrated by the long time it took to get plasmids from other researchers. It could take months before she got the requested plasmids. She saw the need for a central nonprofit repository that would collect, store, do quality control, deal with customs, handle material transfer agreements, and mail plasmids to research labs and other nonprofits. As Addgene's growth shows, Fan was right; a plasmid repository was needed. Before Addgene, post-docs and grad students spent large amounts of time preparing and sending plasmids to other researchers. When Feng Zhang worked in Karl Deisseroth's optogenetics lab at Stanford, he was responsible for sending the channelrhodopsin plasmids to researchers interested in turning on their favorite neurons. Today Zhang is fine-tuning CRISPR-Cas at MIT, but his graduate work/plasmid mailing at Stanford has made him "the most experienced FEDEX label maker at MIT" and led him to deposit his CRISPR-Cas plasmids at Addgene so his students could focus on their research.[12] The first CRISPR plasmids were deposited at Addgene in 2012. Today they account for 20 percent of Addgene's orders. Addgene averages 350 deliveries a day and has sent nearly one million plasmids to more than 80 countries. Scientists can use an online form to request over 50,000 unique plasmids for $65 plus shipping. They can be sent to all countries except North Korea, Iran, Syria, Cuba, and Sudan. To supplement its plasmid repository, Addgene has added educational materials to its plasmid catalog, making available free protocols, blogs, and e-books about the plasmids, CRISPR, and fluorescent proteins. Addgene, with its easy, fast, and cheap distribution of the newest CRISPR-

Cas plasmids, has been a key driver of the technology. "Addgene has played a major role in the exciting democratization of CRISPR, delivering the tools for groundbreaking gene-editing research to investigators in every corner of the globe," says Kevin Davies, executive editor of the *CRISPR Journal*.[13] Jennifer Doudna likes to think of Addgene as the Netflix for molecular biologists.

CRISPR APPLICATIONS

In June 2018, on *Last Week Tonight with John Oliver*, Oliver turned his focus on CRISPR, saying, "One of the most extraordinary aspects of CRISPR-directed gene editing is the broad impact it has had on so many disciplines. . . . It seems at this point that the only thing CRISPR won't be able to do is open your refrigerator door!"[14]

What makes CRISPR such an exciting technique? What can we do with CRISPR that we couldn't do before? Jennifer Doudna has said, "By the summer of 2015, the biotechnology that I'd helped establish only a few years before was growing at a pace that I could not have imagined. And its implications were seismic—not just for the life sciences, but for all life on earth."[15] The short answer to those questions is that genetic modifications that used to take sophisticated biological laboratories years to do can now be done in days. A PhD project costing hundreds of thousands of dollars has become an undergraduate (maybe even high school) project costing a few hundred dollars.

CRISPR has revolutionized molecular biology. Before CRISPR, gene editing was species specific, expensive, and tedious. Now researchers can edit genes in any species, and it's cheap and fast. Thanks to CRISPR, multiple genes can be edited, added, or removed, or have their expression levels modulated at the same time. Not only will CRISPR change what researchers can do in their labs; it will also impact all our lives in one way or another. We have already seen that if you have yogurt for breakfast, the chances are very good that you have eaten cells modified by CRISPR. In the remainder of this chapter I describe some ways in which CRISPR has been applied, show some of its future potential, discuss the ethical dilemmas created by it, and finally return to CRISPR as a WMD.

Plants

On March 18, 2018, the secretary of the USDA, Sonny Perdue, issued a statement that the agency would not regulate genetically edited plant varieties that were indistinguishable from plants that could be obtained by traditional breeding methods.

The domestication of the first agricultural plants can be traced back at least 11,000 years. Since then humans have been using classical plant breeding techniques that rely largely on selective and crossbreeding methods. According to the 2018 statement, the USDA would not regulate potatoes that have been CRISPRed to make fewer carbohydrates, because their existing (potato) genes have been edited. However, cold-resistant tomatoes created by the addition of a fish antifreeze protein would be considered genetically modified plants and would be regulated by the USDA. No matter how good a breeder you are, there is no way a tomato/cold-water fish hybrid would occur naturally. Sonny Perdue's statement led to headlines such as "Crispr'd Food, Coming Soon to a Supermarket Near You"[16] and "USDA Announces Super-Chill Stance on Gene-Edited Crops."[17] CRISPR-modified plants will be produced much more quickly and less expensively than those created by traditional breeding methods and will be indistinguishable from one another. This is true because off-target modifications, which are a problem in animal gene editing, are rare and harmless in plants. Off-target mutations are editing typos that can occur when genes are modified with CRISPR. Although labeling wasn't mentioned in the USDA statement, most experts equate unlabeled with unregulated. Under the USDA definition, "gene edited" is not necessarily "gene modified" and therefore doesn't have to follow GMO labeling laws. In just a year or two we will have many CRISPRed plants in our lives and on our menus; in most cases they will be unlabeled, so we will not even know that they have been created using CRISPR. It is impossible to list all the plants that are being edited to produce variants with improved properties, so I describe here just a few projects that have grabbed my attention, ranging from beer to grass (both kinds).

The University of California, Berkeley, is busy CRISPRing away. Not only is it the home of Jennifer Doudna; it also gave birth to a small start-up called Berkeley Brewing Science. In a 2018 *Nature Communications* paper, the brewers described how they used CRISPR to add two genes to yeast to obtain the hops flavor desired by many beer connoisseurs without actually using hops.[18] Hops require a lot of water and energy to grow and harvest, and a single pint of hops craft beer can require 50 liters of water to grow the hops that give it its distinct characteristics. Berkeley Brewing Science boasts that it has made a sustainable hops-less beer. (Sounds like someone is trying to justify the fun they are having in the lab.) Genes from mint and basil that code for the enzymes that produce their distinctive flavor components were also added to the yeast. In blind taste tests, the hops-less beer revealed no off-taste, and it had notes of Froot Loops and orange blossoms, as well as the desired strong hops flavor. Berkeley Brewing Science wants to market its modified brewer's yeast to other beer brewers. In the future the company hopes to add new nontraditional flavors to its yeasts.

Rice provides one-fifth of the calories consumed by humans. Worldwide production was 741.5 million metric tons in 2014. It is not surprising that scientists are trying to improve rice varieties. Like many of the top scientists today, Professor Jian-Kang Zhu holds many positions. He is a distinguished professor in the Department of Horticulture and Landscape Architecture at Purdue as well as the director of the Shanghai Center for Plant Stress Biology at the Chinese Academy of Sciences. His research team created rice that produces 25–31 percent more grain by using CRISPR-Cas9 to edit 13 genes in the rice.[19] According to Ray A. Bressan, a distinguished professor in the Department of Horticulture and Landscape Architecture at Purdue University and a collaborator of Zhu's, "You couldn't do targeted mutations like that with traditional plant breeding. You'd do random mutations and try to screen out the ones you don't want. It would have taken millions of plants. Basically, it's not feasible. This is a real accomplishment that could not have been done without CRISPR."[20]

Most of the laboratory work described here was done in the United States, while the field tests were done in China. This reflects a modern trend in which scientists working in U.S. and European Union (EU) labs do their field tests and human trials in China or Africa to avoid regulations and stricter safety procedures. Regulating science can be difficult, and like water, it follows the path of least resistance.

Other groups have modified rice, with CRISPR of course, so that it produces more amylose, making a healthier rice in an attempt to mitigate the incidence of type II diabetes, and yet others have created herbicide-resistant rice.

All the major biotech companies, seed companies, and CRISPR start-ups are gene-editing food crops. The big questions are whether these products will be labeled and whether consumers will buy and eat labeled genetically edited products. Zachary Lippman, a geneticist at Cold Spring Harbor Laboratory in New York, might have inadvertently answered that question. On his two acres of experimental crops he has some CRISPRed tomato plants that have more fruit per plant and others that are larger than normal tomatoes. He has brought some of his prize tomatoes home to his wife and six children. Even though they are no different from conventional tomatoes; are considered safe; and have the support of federal regulators in the United States, World Health Organization, European Commission, (British) Royal Society of Medicine, U.S. National Academy of Sciences, and American Medical Association, they have not been eaten. "My wife won't let me," he says.[21] This reminds me of Donna Strickland (chapter 2), who refuses to get laser eye surgery despite having invented the high-intensity lasers used in the surgery.

In the 1950s, breeders found a tomato relative that grew wild in the Galapagos Islands. Its fruit lacked a joint to the stem, making for easy pick-

ing. Breeders crossed this variant with existing commercially grown tomatoes. The resultant tomato variants were perfectly adapted for mechanical
harvesting, but unfortunately when the jointless tomato was crossed with 90
percent of common tomato varieties, the cross had more branches and flowers than normal tomatoes, resulting in fewer and smaller tomatoes. Nevertheless, the easy picking trait was so desirable that breeders persevered with the
new variant and tried to overcome the flower traits by selective and crossbreeding techniques. Now Lippman has found the "jointless" and "flowering" genes, enabling him to use CRISPR to create tomatoes with the advantages of the wild tomatoes without its disadvantages and without the expense
and tediousness of the previously used breeding strategies. [22]

Yi Li, a professor of plant science at the University of Connecticut, is
using his own slightly modified CRISPR technique to reduce the production
of the growth hormone gibberellin in lawn grass, thereby creating a lush
green lawn that requires less frequent mowing. At the same time, he is
researching ways to create lawn grass varieties that require less water and
fertilizer. [23] Li is also applying his CRISPR techniques to create healthier and
less psychedelic strains of "grass." His CRISPRed cannabis plants will have
low levels of the principal psychoactive constituent of cannabis, tetrahydro-
cannabinol (THC), and high levels of the analgesic component, cannabidiol.

Unless the USDA changes its policies and regulates and/or labels CRIS-
PRed plants, I am sure that within the next five years we will be eating gene-
edited products on a daily basis. And like all yogurt-eating folk, we will have
no idea when we are eating gene-edited products, just as we don't know
whether the fruits we eat were obtained by crossbreeding.

Animals

The situation will be different for CRISPRed animals. The FDA has announced that it considers genetic editing of animal DNA a form of gene
therapy and that it would be regulated as a veterinary drug. This means that
we should not expect to be eating meat from gene-edited cows that have no
horns or playing with miniature pigs created for the pet market. Micropigs
and hornless cows have been created by gene-editing (CRISPR and its predecessors), but at this point the FDA prohibits their sale until they have gone
through the FDA regulatory process. This will be a slow and expensive
process, with no guarantee that the successful approvals will be accepted by
the public. Are we ready to eat meat from super muscular, hornless cows?
Does it matter that they have been created by crossbreeding or by gene-
editing? The answer in this case could be yes. CRISPR-Cas9 is great and is
going to change science, but it is not perfect; sometimes off-target edits
occur, and in animals these changes can be dangerous. Fortunately, CRISPR
is fairly specific; it is not going to make many off-target edits, and the few

off-target edits it makes are going to be to sequences that are very similar to the target sequence and therefore predictable.

These off-target edits are particularly important when CRISPR is used to cure genetic diseases in humans. Unlike a pill with an undesired side effect that can be discontinued by no longer taking the pills, gene-editing is permanent, and off-target edits can't be undone. The stakes are higher, and the possibility for error needs to be low if CRISPR treatments are going to become mainstream. To overcome these problems, Feng Zhang and others have created CRISPR-Cas systems that are more accurate and less likely to have off-target edits than the naturally occurring CRISPR-Cas systems. Furthermore, the chances that the off-target edits will hit an important gene and then change it enough to have a substantial effect on the welfare of the patient are slim. As Steven Pinker so eloquently puts it, "Each of us introduces dozens of mutations into our own germlines by exposing ourselves to everyday radiation and chemical mutagens. Genetic editing would be a droplet in the maelstrom of naturally churning genomes."[24]

More than any other new technology, CRISPR is opening doors to new experiments, organisms, and ethical dilemmas. CRISPR could be used to create larger and more nutritious crops, flowers with a riot of new colors, superbly muscled livestock, mice models fine-tuned for the study of human diseases, and mini pigs for the pet trade; it could also be used to wipe out whole species of organisms and enhance human embryos. Should we regulate these experiments? If yes, how? Where do we draw the line? Jennifer Doudna has done a lot of thinking about CRISPR. She doesn't have a solution, but I like her ideas, many of which have been expressed in her book, *A Crack in Creation: Gene Editing and the Unthinkable Power to Control Evolution*, in which she writes, "I hope that gene edited livestock will make agriculture more humane and environmentally friendly, not just profitable. . . . I hope that a shared respect for animal welfare will temper these and similar efforts."[25]

For now, let us review some of the interesting CRISPR animal applications, remembering that most of these experiments are in their very earliest stages, and in some cases the experiments have been done using gene-editing techniques that predate CRISPR (they are from the BC age, i.e., Before CRISPR, e.g., a technique called TALEN). For all of these it will be difficult to get gene-edited animals and animal products to market in the United States.

In 2017, Jianguo Zhao of the Institute of Zoology at the Chinese Academy of Sciences in Beijing led a team of researchers who used CRISPR to add the gene responsible for burning fat to regulate body temperature in mice to pigs. Pigs normally lack this gene, which probably disappeared from the modern pig line about 20 million years ago. The resultant genetically modified pigs had 24 percent less fat and were less sensitive to cold.[26] In colder climes, heat lamps are often used to keep pigs warm; these GMO pigs would require less heating in the winter. They would also be cheaper to rear than normal pigs, because fatter pig breeds eat more and grow more slowly. Although the gene is found in many mammals besides mice, and the resultant pork will be just as healthy as that obtained from non-GMO pigs, it is unlikely that these pigs will be seen/eaten in the United States or EU due to public opinion about GMOs and livestock regulations. However, they may make it onto the Chinese market.

Belgian blue cows were created in the 19th century by crossbreeding cattle from northern and central Belgium with British shorthorns. Some Belgian blues are super muscular; they look as if they have been on a diet of steroids. They have 20 percent more muscle than typical cows, have a larger proportion of desirable cuts of meat and less fat. The bulls can reach a weight of 1,250 kg, and the cows weigh in at about 900 kg. These bovine body builders have something called double muscling. Double muscling was first recorded in 1807 and is also observed in Piedmontese cattle. In 1997, researchers from the USDA Meat Animal Research Center in Nebraska and AgResearch in New Zealand determined that a single gene, myostatin, was responsible for the double muscling. In Belgian blues the gene is missing 11 nucleotide bases, and in Piedmontese there is a single-letter mutation.[27] The myostatin gene curbs the body's production of muscle tissue; without myostatin or with impaired myostatin, excessive muscle tissue is produced.

The quality of life of these double-muscled cows is low. In 90 percent of their births a cesarean section is the only option. These supercows have joint problems, low stress tolerance, and enlarged tongues. The Danish government, PETA, and other animal rights organizations have asked breeders to stop breeding double-muscled cattle.

However, groups all over the world are very interested in using CRISPR to create double-muscled livestock. The main reason for this is that double muscling involves editing just one gene, myostatin, that is found in most livestock. No foreign genes are added, so the gene edits could be obtained by traditional methods, which in turn, in some countries, could mean that there is less regulatory oversight. In addition, double-muscled livestock will lead to increased meat production and improved food security.

In 2015, Bruce Whitelaw from the Roslin Institute (birthplace of Dolly the cloned sheep) in Edinburgh, Scott Fahrenkrug from Recombinetics Inc. in St. Paul, Minnesota, and their collaborators edited the myostatin gene in Nellore cows. Using gene-editing, they fast-tracked conventional breeding techniques so that Nellore cows also produced double-muscled offspring.[28] Since then, double-muscled sheep, pigs, goats, rabbits, and dogs (beagles) have been created with edits to the myostatin gene. And, yes, as we saw in chapter 3, Josiah Zayner tried to inhibit the myostatin gene on his own arm.

Prior to the advent of CRISPR, mice were the most popular mammalian model organisms for studies of human diseases and conditions, in part because they could be gene edited with the techniques that existed before CRISPR. More than 30,000 distinct mouse varieties are available. In many cases they are excellent models for studying human diseases; however, for some diseases, such as cystic fibrosis and neurodegenerative diseases, mice aren't good models for humans. For these diseases, pigs are much better model organisms. Since we have CRISPR, we can gene-edit pigs as easily as we can mice. Our curly tailed counterparts also have organs similar in size to humans, have large litters, have short gestation periods, and can be kept in pens, making them popular but significantly more expensive alternative model organisms to mice.

Pigs can be more than model organisms. They can also be the source of organs for transplants into humans. A heart transplant is the final option for patients with severe coronary artery disease and end-stage heart failure. Christiaan Barnard performed the world's first heart transplant in 1967 in Cape Town, South Africa; the patient died of pneumonia less than a year after the transplant. Since then, the surgical procedure has been improved, and the average post–heart transplant survival rate is 15 years, with more than 70 percent of patients who undergo a heart transplant surviving for more than 5 years. Thanks to new alternative treatments and the high price of a heart transplant, about $750,000 in the United States, there is a diminishing demand for heart transplants; nevertheless, there are not nearly enough donors to meet the demand. About 3,500 heart transplants were performed worldwide in 2007, while 800,000 people have heart conditions that warrant a transplant.

There were more than 114,000 people awaiting organ transplants in the United States in 2018, and there is an urgent need to supply them with replacement organs that is not being met by human donations. According to the FDA, every day about 20 people die while on a waiting list for a vital organ transplant.[29] There are two alternatives to heart transplants: the use of artificial hearts or hearts from another species (xenotransplantation).

Although primates are our closest relatives, they cannot be bred in captivity in large numbers, and because it is thought that the human immunodeficiency virus (HIV) may have developed from its primate analog, the simian immunodeficiency virus (SIV), there is a fear that primate organs could spread HIV. Pigs are the next best source of organs; their organs are roughly the same size as their human counterparts', and they are evolutionarily distant enough that the two species do not share many pathogens. In order to make porcine transplants a reality, two major obstacles have to be overcome: rejection and disease transmission. In the late 1990s, pharmaceutical companies took notice of the possibilities of transplants from pigs. Novartis planned to invest more than $1 billion in xenotransplantation. PPL Therapeutics (the creators of Dolly the sheep) and Genzyme also invested heavily. But by the early 2000s it became apparent the pig genome was littered with porcine endogenous retrovirus (PERV) genes and that the infection and rejection problems were complex and wouldn't be solved with the currently existing research methods. Big pharma research into porcine xenotransplantation stopped. The advent of CRISPR changed that situation. George Church, a rather ambitious geneticist at Harvard, who was one of the first researchers to use CRISPR to gene-edit human cells in 2013, saw porcine xenotransplantation as the perfect CRISPR challenge. In a scientific tour de force, he marshaled his laboratory to use CRISPR to remove all 62 PERVs in pig cells,[30] a feat that would have been impossible without CRISPR. In 2015, Church and Luhan Yang, a former student of his, parlayed these results into the founding of a new xenotransplantation biotech company, eGenesis. It has raised over $38 million in start-up funds and employs at least 12 former members of the Church research group.

One of the first results to emerge from a transnational (U.S. [Harvard and eGenesis], China, and Denmark) collaboration was a *Science* paper demonstrating that PERVs do indeed infect human cells, which can in turn infect other human cells, and perhaps equally important, that the researchers had created 37 live, healthy PERV-free pigs in a trial.[31] Having solved the problem of transplanted pig organs infecting their human hosts with PERVs, eGenesis now has to solve the rejection problem. "Trying the straight transplant failed almost immediately, within hours, because there's a huge mismatch in the carbohydrates on the surface of the cells, in particular alpha13galactose, and so that was a showstopper," Church explains. "When you delete that gene, which you can do with conventional methods, you still get pretty fast rejection, because there are a lot of other aspects that are incompatible. You have to take care of each of them, and not all of them are just about removing things—some of them you have to humanize. There's a great deal of subtlety involved so that you get normal pig embryogenesis but not rejection."[32] Although rejection is a substantial obstacle, eGenesis believes CRISPR is up to the challenge.

George Church is on many boards and has cofounded 22 companies. One of his more interesting endeavors is the Harvard Woolly Mammoth Revival project. Church would like to use CRISPR to de-extinct (yep, this also seems to be a word) the woolly mammoth. Scientists have partially sequenced the mammoth genome from DNA obtained from old mammoth bones. Beth Shapiro is a professor of ecology and evolutionary biology at the University of California, Santa Cruz, and the author of the book *How to Clone a Mammoth*. In her book she describes the difficulty in sequencing the genetic material of extinct animals. When an animal such as a woolly mammoth dies, the DNA degrades, leaving scientists with what Shapiro describes as "a soup of trillions of tiny fragments" that require reassembly.[33] There is not enough information in that soup to fully re-create a woolly mammoth today, leaving Church and his researchers with no option but to use the genetic information they have to modify Asian elephants, which are the woolly mammoths' closest living relatives, so that they become modern hybrids that look and behave like mammoths. So far the Harvard Woolly Mammoth Revival project has successfully incorporated selected mammoth genes into skin cells of Asian elephants, marking the first time that woolly mammoth genes have been functionally active since the species became extinct.

Church's Harvard Woolly Mammoth Revival project isn't the only de-extinction game in town. Stuart Brand, the creator of the *Whole Earth Catalog*, has cofounded the conservation group Revive and Restore, which is working on reviving the extinct passenger pigeon and the heath hen. As you can well imagine, these de-extinction projects have a polarizing effect on public opinion. Many detractors, myself included, see no reason why we should spend all this effort to create poor copies of long-dead species when there are plenty of modern-day species in dire need of conservation. Other critics see this as "unnatural and even an unnecessary show of scientific hubris."[34]

Medicine

There are more than 10,000 genetic disorders caused by mutations that occur on only one gene, the so-called single-gene disorders. They affect millions of people. Sickle cell anemia, cystic fibrosis, and Huntington's disease are among the most well-known of these disorders. These are all obvious targets for CRISPR therapy.

"The pace of progress in gene therapy has been somewhat breathtaking," Scott Gottlieb, the FDA commissioner, said in May 2018.[35] A publication from MIT[36] predicted that there would be about 40 CRISPR-based gene therapy products approved by the FDA by the end of 2022. It also predicted that 45 percent of these approvals would be for products targeting cancer. Here I very briefly introduce just a few medicinal uses of CRISPR.

Editas Medicine, a pharmaceutical company founded by Feng Zhang, George Church, and David Liu, is using a CRISPR-based therapy to treat a hereditary form of blindness, Leber congenital amaurosis, which is a genetic disorder that causes serious vision loss. In monkeys the therapy can restore light sensitivity in 10 percent of defective photoreceptor cells. The company has FDA approval for human trials with up to 18 patients. It is the first company to get permission to do human in vivo CRISPR studies.

Hemoglobin, a critical protein in red blood cells, transports oxygen and carbon dioxide around the body. Beta thalassemia and sickle cell anemia, two human disorders that are caused by mutations in a hemoglobin gene, are perfect targets for CRISPR therapy. More than four million people have sickle cell disease. Their average life expectancy is 40 to 60 years. Sickle cell disease is responsible for repeated bouts of pain, swollen hands and feet, and stroke. About 1 in 100,000 people in the world have beta thalassemia. Children in the first two years of life display symptoms that include skeletal abnormalities, slow growth, and anemia. Untreated beta thalassemia can lead to heart failure. CRISPR treatments for sickle cell anemia and beta thalassemia have been tested in monkeys. In both cases red blood cells are removed from the monkeys, gene edited with CRISPR to correct the genetic mutations, and returned back to the monkeys (ex vivo treatment).

Hans-Peter Kiem at the Fred Hutchinson Cancer Research Center has gene edited stem cells found in monkey blood. Forty percent of the cells were successfully edited, the changes lasted for more than six months, and no off-target edits were observed. "Since monkeys are so similar to humans, I don't think there's going to be a huge challenge in translating this work to humans," Kiem said in an interview with *MIT Technology Review*. "We use the same technology as we would in patients."[37]

Presently beta thalassemia and sickle cell anemia in humans are treated by bone marrow transplantation from donors who have immunologically matching cells. There are nowhere near enough donors willing to undergo the invasive procedure. CRISPR has the dual advantage that the patient is both the donor and the host; there will be no rejection problems, and the patient is very invested in undergoing the procedure. Numerous academic labs and biotech companies are racing to be the first to use CRISPR-edited stem cells in ex vivo treatment of beta thalassemia and sickle cell anemia. In November 2019, the early results of the first human trials (fewer than ten patients) were announced. They were encouraging for both sickle call anemia (CRISPR Therapeutics and Vertex Pharmaceuticals) and beta thalassemia (Bluebird).

Ex vivo treatments, in which cells are removed from the patient, gene edited, checked for off-target edits and other problems, then returned into the host, are much simpler and safer than in vivo treatments. However, most diseases will have to be treated by in vivo methods.

One of the ways to overcome in vivo delivery problems is by genetically modifying organisms when they are in their embryonic stage, so that all the cells arising from the genetically modified embryonic stem cells will be modified.

Hypertrophic cardiomyopathy is caused by an MYBPC3 mutation that is found in about 1 in 500 adults and leads to heart failure. It is the most common cause of death in otherwise healthy young athletes. Current treatments can offer symptomatic relief but can't cure the underlying genetic causes. In August 2017, Shoukhrat Mitalipov and his colleagues at Oregon Health and Science University published their work describing how they had edited the MYBPC3 gene in human embryos that were then not allowed to develop beyond the blastocyst stage.[38] The paper has 30 coauthors from four countries; that is common in today's science. In their study, Mitalipov and colleagues used clinical-quality human eggs donated by healthy women and semen from an adult male patient with a familial history of hypertrophic cardiomyopathy. By editing an embryo in the very early stages of development, the genetic changes made were incorporated into all cells derived from the edited cells, including germ cells (the only cell type capable of generating an entirely new organism). "Every generation on would carry this repair because we've removed the disease-causing gene variant from that family's lineage," says Mitalipov. "By using this technique, it's possible to reduce the burden of this heritable disease on the family and eventually the human population."[39] In more than half the embryos the MYBPC3 mutation was fixed and the embryos grew to the blastocyst stage, in which they could be transferred back into the a woman's uterus if this was part of an in vitro fertilization (IVF) procedure. This was the first time CRISPR had been used to genetically modify a viable embryo. If the embryos hadn't been destroyed during their analysis, they could have been used to produce IVF babies free of hypertrophic cardiomyopathy. Mitalipov and his colleagues didn't implant the modified embryos because they didn't know whether there would be CRISPR side effects or whether the resultant babies would be healthy and because there were no ethical guidelines to guide them from experiment to treatment. Little did they know that on the other side of the world researchers were ignoring ethical and safety considerations.

The scientific community got an unpleasant surprise in November 2018 when He Jiankui, an associate professor at the Southern University of Science and Technology in Shenzhen, China, took the next step. He announced that he had used CRISPR-Cas9 to genetically modify twin girls, named Lulu and Nana, when they were still in the embryonic stage. Jennifer Doudna was "horrified" by the announcement, and NIH director Francis Collins said the experiment was "profoundly disturbing"; many other echoed these opinions.[40] George Church was the only prominent scientist to come to He's defense.[41] Fertility clinics were not so reticent; within a week after his announcement He Jiankui received an email from a Dubai fertility clinic, congratulating him and expressing interest in collaborating. More requests would follow. There are many fertility clinics eager to CRISPR in vitro embryos, and there are bound to be many clients rich and foolish enough to pay them for the privilege.[42]

The main concern that most scientists have with the work is that the twins were not CRISPRed to meet an unmet medical need. He Jiankui deactivated a gene called CCR5 to reduce the risk that they would get HIV. They didn't have HIV and most likely would never have gotten HIV. Lulu and Nana were genetically modified as a proof of concept experiment. Leading AIDS researcher Anthony Fauci was quick to point out that "there are so many ways to adequately, efficiently, and definitively protect yourself against HIV that the thought of editing the genes of an embryo to get to an effect that you could easily do in so many other ways in my mind is unethical."[43]

Furthermore, many experts in the field have said that the experiments were amateurish. He Jiankui did not follow standard protocols, he never got the correct consent forms from the parents of the girls, and he didn't get permission from the appropriate ethical committees at his university; instead, he took an unpaid leave so that he could work without interference. Although Professor He consulted with faculty he knew at Rice, Berkeley, and Stanford, he didn't take their advice; nor did they do anything to stop him. He Jiankui never published his results. Instead he used an American public relations consultant to orchestrate his announcement so that it coincided with the release of a series of five YouTube videos. In an article in the *Atlantic*, Ed Yong lists 15 concerns he has with He's experiments and writes, "If you wanted to create the worst possible scenario for introducing the first gene-edited babies into the world, it is difficult to imagine how you could improve on this 15-part farce."[44]

Alcino Silva, a neuroscientist at the University of California, Los Angeles, who does research on the CCR5 gene, thinks that disabling the gene is akin to removing the brakes on a car. "The car would go a lot faster," Silva says, but the risk of harm would be higher. "Evolution has worked hard," he says, "to give us the genes that we need."[45] This was not a good target gene. To add insult to injury, a paper published in *Nature Medicine* in June 2019 reported that people with two disabled copies of the CCR5 gene are 21 percent more likely to die before the age of 76 than are people with at least one working copy of the gene.[46] He Jiankui's genetic modifications may in the end shorten the lives of Lulu and Nana.

The Chinese regulatory authorities, who are getting gene editing research into human trials faster than any other country in the world,[47] reacted quickly and stopped Jiankui from doing further research on human embryos. By making heritable mutations (see the germline discussion later in this chapter), He had crossed the line. China does not want to rock the boat; it has a lot invested in gene editing. It would like to maintain credibility among the scientific community so that it can continue to offer its labs and facilities to foreign researchers and outsource its expertise. On December 30, 2019, the Shenzhen Nanshan District People's Court sentenced Jankui to a three-year prison term and fined him $425,000. He and two of his colleagues were sentenced for gene-editing human embryos and deliberately violating the relevant national regulations on scientific research and medical management. The court statement mentioned a third gene-edited baby born in the summer of 2019 for the first time.[48]

GENE DRIVES

Gene drives force a genetic trait through a population, defying the usual rules of inheritance. Normally there is roughly a 50 percent chance that a genetic trait will be passed from a given parent to the offspring; with a gene drive that possibility can become a near certainty. In proof of concept experiments, yellow fruit flies with a gene drive associated with their coloring were released in a caged population of wild type fruit flies. Within a few generations, 97 percent of the fruit flies were yellow, and no wild type yellow-brown fruit flies remained. It has been estimated that if one yellow fruit fly had escaped the lab, between 20 and 50 percent of all the world's fruit flies would now be yellow. That is a powerful technique.

With the combination of CRISPR and gene drives, we have overcome constraints imposed by nature. We are no longer limited by the rules of inheritance. Evolution is in our sights and in our hands; using CRISPR we can control a genetically defined trait and push it through a population with a gene drive. This is an incredibly powerful technique that has the potential to save millions of lives (and to wipe out an entire species).

Climate change and international travel are responsible for the tripling of insect-borne diseases in the United States that occurred between 2004 and 2016. The chief culprits are ticks, mosquitoes, and fleas. Globally, mosquitoes are doing most of the damage; they spread malaria, Zika, and dengue and yellow fever. In the global North countries the insecticide DDT was used to eradicate *Anopheles* (malaria carriers) and *Aedes* (dengue) mosquitoes, but with climate change and the discontinued use of DDT, they are coming back to Southern Europe and the United States. Most people who get malaria have debilitating headaches and fever but survive. About 200 million people are infected each year. According to the World Health Organization, "Malaria disproportionately affects poor people, with almost 60% of malaria cases occurring among the poorest 20% of the world's population."[49] In sub-Saharan Africa, more than 300,000 people are killed annually by malaria, and 75 percent of the victims are children under age five. Not only is this part of the world the home of the most lethal malaria parasite of them all, *Plasmodium falciparum*, but the parasite's host, *Anopheles gambiae*, is a vicious disease vector. This mosquito does not hibernate, is the most aggressive of all mosquito species that are capable of harboring *Plasmodium*, and is the only mosquito to get its blood meals exclusively from humans.

Andrea Crisanti of Imperial College London has spent 15 years and about $100 million in funding to create gene drives that interfere with mosquitoes' sex chromosomes. Now he can collapse a caged population of *Anopheles gambiae*. "This is the first time we have shown that we can in principle manipulate the fate of an entire species," says Crisanti, whose groundbreaking work has been supported in large part by the Bill and Melinda Gates

Foundation.[50] Crisanti has teamed up with institutions in Burkina Faso, Uganda, and Mali and hopes to release his self-destructive mosquitoes in the wild in 2024. One of the main advantages of using gene drives to collapse a population of disease-carrying insects, such as *Anopheles gambiae*, is that it is essentially a species-specific pesticide. The CRISPRed mosquitoes will only mate with other *Anopheles gambiae* and won't be interested in any other of the hundreds of mosquito species. One can start a genetic war with the malaria-carrying mosquitoes without affecting other mosquitoes, bees, butterflies, and so forth.

In a variation of the Crisanti experiments, other research groups are using gene drives and CRISPR as genetic guerrilla fighters designed to alter the mosquitoes to make them and all their offspring inhospitable to the malaria parasite. Although they will no longer be hosts to malaria parasites, these *Anopheles* mosquitoes will live normal, healthy lives and won't be wiped out.

Kevin Esvelt leads the Sculpting Evolution Group at the MIT Media Lab. In 2013, he was the first to recognize the potential of combining CRISPR and gene drives to alter the genetics of a wild population of organisms. He is one of the most thoughtful researchers in the gene drive field. From the beginning he has been frightened by the potential of his own findings and has called for open discussions and safeguards before releasing transgenic organisms with gene drives in the field. Esvelt's opinions reflect those of most working in the field and have been expressed at a number of conferences and governmental panels convened to examine gene drives: despite the potential risks of unregulated widespread genetic manipulation, gene drive research needs to continue because the potential benefits are too great to institute a research moratorium. However, we haven't yet reached the point where we can release the products of our research into the wild. "There's always a cost to doing nothing, and we need technology not just to keep the world running, but also to make it better," Esvelt told CNN in 2018.[51]

Lyme disease originates from Lyme bacteria, found in the blood of white-footed mice. Ticks carry the bacteria to deer and humans. "Mice against ticks" is an Esvelt project designed to eradicate Lyme disease on Nantucket and Martha's Vineyard. None of the research is carried out in the field; nevertheless, Esvelt is already reaching out to the communities living on the two islands. Science outreach and education is part of his job. He hopes that in five years he and the community will be ready for the release of 1,000 genetically modified mice. In Esvelt's view, using gene drives is risky, but less so than alternatives. "Better that we use DNA than potentially inhumane pesticides," he says.[52]

Gene drives have tremendous potential for good, and unfortunately just as much potential for bad. It should not be surprising that the largest single funder of gene drive technology is the U.S. Defense Advanced Research

Projects Agency (DARPA). It has spent approximately $100 million on its "safe genes" project.[53] On its website DARPA describes the program thus: "The Safe Genes program supports force protection and military health and readiness by protecting Service members from accidental or intentional misuse of genome editing technologies. Additional work will leverage advances in gene editing technology to expedite development of advanced prophylactic and therapeutic treatments against gene editors. Advances within the program will ensure the United States remains at the vanguard of the broadly accessible and rapidly progressing field of genome editing."[54] DARPA is concerned that gene drives could be weaponized to target food sources or the human microbiome.

SHOULD WE BE DOING THIS?

In a 1967 *Science* editorial, Marshall Nirenberg, who won the 1968 Nobel Prize in medicine for establishing how DNA gets translated into proteins, said,

> Man may be able to program his own cells with synthetic information long before he will be able to assess adequately the long-term consequences of such alterations, long before he will be able to formulate goals, and long before he can resolve the ethical and moral problems which will be raised. When man becomes capable of instructing his own cells, he must refrain from doing so until he has sufficient wisdom to use this knowledge for the benefit of mankind. I state this problem well in advance of the need to resolve it, because decisions concerning the application of this knowledge must ultimately be made by society, and only an informed society can make such decisions wisely.[55]

Thanks to Jennifer Doudna and many others, that time has come; we are capable of instructing our own cells. However, I am afraid society is not well enough informed to make wise decisions about human (or plant or animal) genetic engineering, and we don't have sufficient wisdom to use CRISPR for the benefit of humankind. "What will we, a fractious species whose members can't agree on much, choose to do with this awesome power?," asks Jennifer Doudna.[56]

My three main areas of concern are using gene drives in the field, using CRISPR to modify germ cells so that future generations inherit the genetic modifications, and CRISPRing humans to improve their nonmedical needs: their abilities and looks. Kevin Esvelt thinks about gene drives a lot. He admits, "My greatest fear is that something terrible will happen before something wonderful happens. It keeps me up at night more than I would like to admit."[57]

The dangers of using CRISPR and gene drives in disease vectors such as mosquitoes and ticks aren't fully understood, thus making it difficult to calculate the risks. Furthermore, regulating gene drives and getting consent for releases of genetically modified organisms is complicated by the fact that these organisms don't respect municipal and national borders. Gene drives are self-sustaining; they are not like applying a pesticide, and once the drive starts propagating through a population, there is no easy way to stop it. Toward this end, many groups are working on reverse drives, molecular emergency brakes that are CRISPRed into the organism with the gene drive.

Thanks to gene drives, we have gained evolutionary power over nature. This isn't very reassuring given our record of looking after nature.

Human germline editing is the process by which the genome of an individual is edited in such a way that the change is heritable. This is achieved through genetic alterations within the reproductive cells, that is, the egg and sperm. One might argue that we have a lesser responsibility when genetically modifying ourselves than when we modify other organisms. But it is important to remember that the genetically altered offspring and all their subsequent offspring don't have a choice in the matter. We would be deciding for all future generations. There are very few diseases in which germline modifications are the only way to make sure that children will be born without the disease, such as sickle cell disease, Huntington's, Marfan syndrome, cystic fibrosis, and Fanconi anemia. In most other genetic diseases there is a risk, but not a certainty, that the children will get the disease. This makes it difficult to decide when germline editing should be allowed.

Putting rules in place to regulate germline editing is a difficult procedure. Countries have to consider safety, ethics, and competitiveness. If the regulations are too strict, countries will lose their competitive edge as researchers move abroad or do their experiments in secret, as He Jiankui did. This is the worst possible outcome. In addition, CRISPR is so inexpensive and easy to use that it can be done by amateurs in their garages, making enforcement of regulations an even more complex procedure. And even if regulations are put in place, CRISPR tourists will just get their treatments in clinics located in countries with relaxed gene editing laws and less stringent safety regulations.

Fortunately genetic engineering is in its infancy, and I like to think it is still possible to regulate it and prevent future heritable modifications caused by germline editing from being allowed. Currently rules about germline editing vary by country. Canada, Germany, France, Brazil, and Australia have forbidden germline modifications and have legislated punishments ranging from significant fines to jail sentences. In China, Japan, and India, germline modifications are forbidden, but the prohibitions haven't been legislated and are therefore less enforceable. In the United States, government agencies have spoken out about clinical uses of germline modifications, but there are no bans or legislation in place.[58] It would certainly help to have some global guidelines. In the concluding chapter I present an interesting model that might serve as a template: the International Accounting Standards Foundation.

Ethical considerations of germline editing are further complicated by the fact that religions have differing views of the embryo. Jewish and Muslim traditions do not consider embryos created outside the womb to be people and are more tolerant of experimentation with embryos, while many Christian sects oppose germline modifications because they consider the embryo a person from the moment conception has occurred.

Many genetic diseases are associated with tremendous suffering, and it will be very difficult to say no to CRISPRing these patients, especially if this can be done without affecting future generations (as in the case of somatic cell editing rather than germ cell editing). But we also have to acknowledge that once we have allowed gene editing for a defined medical need, the techniques and knowledge gained from these processes will be used to genetically enhance embryos and perhaps even grown individuals. In *A Crack in Creation*, Jennifer Doudna lists a few simple mutations that might tempt future parents. Mutations in the EPOR gene confer increased endurance, mutations in the LRP5 gene are responsible for extra-strong bones, changes in DEC2 lower the sleep requirements of people, MSTN controls muscle growth (it is the myostatin gene Josiah Zayner tried to CRISPR), and finally, my favorite, mutations in the ABCC11 genes are associated with lower levels of armpit odor and the type of earwax produced. As the possibility of CRISPR-designed babies becomes more real, the search to find more simple enhancements will accelerate. And genetics, like all of science, is growing faster and faster. It took seven years to sequence the first 1 percent of the human genome, then just seven more years to sequence the remaining 99 percent. By the time you read this book, the list of potential improvements will be longer.

New products, pharmaceuticals, and medicines always cost more and are therefore more accessible to the rich. This will obviously also be the case for medically needed and enhancement-driven human genetic editing. We need only look at IVF treatments, which still cost tens of thousands of dollars four decades after the first IVF baby, Louise Brown, was born.

Yuval Harari, Stephen Hawking, and Jennifer Doudna have argued, and I agree with them, that gene editing is the same as IVF yet different in an important way. It will widen existing wealth inequities, resulting in genetic inequities and creating something Doudna calls the "gene gap." This is the first time in history that not only will the rich have better lives, but their offspring will actually have the opportunity to be better people with stronger bones, more endurance, and shorter sleep requirements. Of course, like cell phones, they might be out of date before they know it, as newer, improved, younger generations have the newest and best genetic operating system.

There is also the distinct possibility that not only will gene editing magnify existing inequities, but the rich will become less diverse as they edit out all perceived disabilities, including but not limited to deafness, short-sightedness, and obesity. Eventually, they will all have the same fashionable upgrades. Will these changes lead to increased discrimination against differently abled people? When will we reach the point where our genetically enhanced offspring are no longer human?

In *Falter: Has the Human Game Begun to Play Itself Out?*, Bill McKibben worries that genetic editing will wash out the meaning of things "as peripheral as sports. . . . I don't think I'd bother going if stock car races were run by driverless cars. They could doubtless go faster, just as runners genetically altered to have more red blood cells can doubtless go faster. But faster isn't really the point. The story is the point. If something as marginal (though wonderful) as sports can see meaning leach away when we mess with people's bodies or remove them from the picture, perhaps we should think long and hard about more important kinds of meaning. The human game, after all, requires us to be human."[59] Perhaps now we can see why Jennifer Doudna thinks that "with our mastery over the code of life comes a level of responsibility for which we, as individuals and as a species, are woefully unprepared."[60]

CONCLUSION

In concluding this chapter, let us return to James Clapper, the U.S. director of national intelligence, who included gene editing among the six most threatening WMDs. He did this because "research in genome editing conducted by countries with different regulatory or ethical standards than those of Western countries probably increases the risk of the creation of potentially harmful biological agents or products. Given the broad distribution, low cost, and accelerated pace of development of this dual-use technology, its deliberate or unintentional misuse might lead to far-reaching economic and national security implications."[61] This may be ironic given the lax regulations on genetic editing in the United States, but it must be taken seriously. CRISPR has the potential to change our lives for better and for worse. But we have to remember the bigger picture: CRISPR is an extreme example of science in action; its growth is faster and its potential uses and misuses greater than other areas in science. It rightly demands our attention. And although some parts of it are certainly representative of science today (e.g., large multinational, multidisciplinary teams working in competition), we have to remember that the majority of science is not as glamorous. It doesn't appear in our news headlines as often, is less dangerous and less controversial, and yet it remains equally important.

Part Five

Bad Science

All that ridiculous information can make you wonder is science bullshit, to which the answer is clearly no, but there is a lot of bullshit currently masquerading as science.—John Oliver, *Last Week Tonight: Scientific Studies*, HBO, May 16, 2016

Science itself is never inherently bad or evil. However, it is easy to do sloppy science, misrepresent science, or misuse science, and that is what I would call bad science. As science grows bigger and more powerful, these abuses escalate proportionally. This leads to mistrust of science at just the time when it is most important for scientists and nonscientists to find a consensus on sensitive issues, including ways in which we can address climate change and regulate gene editing. It is a selective mistrust of science; most science deniers still consult doctors when they are sick and trust that planes won't fall out of the sky and elevators will stop at the desired floor. Where does the lack of trust in science come from? I believe the main culprits are the systematic undermining of science by the tobacco industry, the inherent complexity of science, the growth of fake news, excess hype on social media, charlatans and quacks who profit from our inability to distinguish pseudoscience from real science, and the politicization of science.

Chapter Ten

This Is Not Science; It Is Fake Science

Freedom of opinion is a farce unless factual information is guaranteed and the facts themselves are not in dispute. Conceptually, we may call truth what we cannot change; metaphorically, it is the ground on which we stand and the sky that stretches above us.—Hannah Arendt, "Crises of the Republic"

Truth and facts are central to the workings of science. We can't solve nature's puzzles if we are given false puzzle pieces. The emergence of fake news and post-truth has undermined the scientific process and trust in science. Where did they come from?

TOBACCO'S WAR ON SCIENCE

In 1761, John Hill published a paper titled "Cautions against the Immoderate Use of Snuff." In it he linked the tobacco found in snuff to lip, mouth, and throat cancer. The ideas and thoughts he expressed in the pamphlet were correct; however, it would take a few hundred years for his findings to be generally accepted.

Fifteen years after Hill described the link between tobacco and throat cancer, Percival Pott, a surgeon at St. Bartholomew's Hospital, London, noted that chimney sweeps had an excessively high rate of scrotal cancer. He attributed the formation of the scrotal cancer to exposure to soot. Hill and Pott were among the first to identify a cancer-causing agent, a carcinogen. During Pott's lifetime, children as young as five years old were apprenticed as chimney sweeps. Pott died in 1788, and in that year the Chimney-sweepers Act was passed, thanks in part to his advocacy about the dangers of the job. The act forbade master sweeps from having apprentices younger than eight years old. It took a few decades after Pott's manuscript was published for chimney sweeps to become aware of the dangers of soot and learn how to avoid it, especially in their scrotal areas. Once this knowledge caught on, the incidence of scrotal cancer for chimney sweeps decreased. Unfortunately, the same can't be said for Hill's work; centuries after he proved the link between tobacco use and cancer, we are still trying to combat lung cancer by reducing smoking. Why is that?

Clearly nicotine's addictive powers have contributed, but equally important is the tobacco industry's continual denial of the dangers associated with smoking. For example, in 1953 the tobacco industry formed the Council for Tobacco Research (CTR) to cloud the public debate about tobacco and health. Its January 1954 "Frank Statement to Cigarette Smokers" reached approximately 43 million people through more than 400 newspapers in the United States. It was drafted by the leading public relations firm of the time with the explicit aim of instilling doubt about scientific research that linked tobacco smoking with disease; it was a forerunner of today's fake news. "For more than 300 years tobacco has given solace, relaxation and enjoyment to mankind," part of the ad read. "Critics have held it responsible for practically every disease of the human body. One by one these charges have been abandoned for lack of evidence."

"Doubt is our product," a tobacco executive wrote in 1969 in an internal memo.[1] In order to undermine scientific research that indicated a link between nicotine and lung cancer, the tobacco industry systematically challenged scientific evidence and the way science was done. It used sophisticated public relations approaches to challenge and distort the science that proved smoking was carcinogenic. Together with a team of advertising agencies and lobbyists, the tobacco companies managed to get politicians and journalists to accept that all science, particularly that surrounding tobacco, was never 100 percent certain, that a "balanced" perspective with views from both sides was always needed, and that inaction was the wisest response to uncertainty.

The lobbying was so effective that in 1962, when U.S. surgeon general Luther L. Terry established an advisory committee on smoking, he ensured that the committee was "unbiased," which meant that he gave the tobacco industry veto power over anyone on the committee and the power to nominate committee members.

Today it would be difficult to find someone who doesn't believe there is a link between tobacco smoking and lung cancer, but the damage has been done. The general public is still skeptical of science; "balanced opinions" are still presented even when the science is solid; and the tobacco lobby's tactics are used by other groups that want to dispute scientific evidence, such as anti-vaccine groups and climate change deniers, not to mention the tobacco industry's resurgence of these strategies with the advent of vapes and electronic cigarettes as a "healthier" alternative to smoking cigarettes.

In "Consilience and Consensus," science writer and historian Michael Shermer, describes why scientific theories are an easy target.[2] All scientific theories start out as an idea proposed and believed by a small minority of scientists. If the theory is valid and robust, the evidence collected from multiple lines of research all converges on the same conclusion, which leads to the acceptance of the theory as scientific truth. In the case of tobacco, climate change, and vaccination science, the convergence has long been reached. However, skeptics and industries with a vested interest (e.g., tobacco and energy companies) have attacked many strands of research in efforts to claim that convergence hasn't been attained yet. They don't have to disprove the theory by unraveling the convergence; all they need to do is find some experiments that are easy to manipulate, weak, controversial, or hard to understand, and pick on them. By focusing on the weakest link, which is given media exposure equal to that of the hundreds, if not thousands, of experiments that prove the theory, the deniers instill doubt.

A 1999 antiracketeering case brought by the U.S. Department of Justice against seven tobacco companies gave the public its first glimpse of the manipulations undertaken by the CTR. Among the records released were documents showing their multipronged approach to co-opting scientists and science. Mark Petticrew and Kelley Lee of the London School of Hygiene and Tropical Medicine used the documents to try to understand the influences of corporate tobacco policies on public health, particularly in the realm of stress research.[3]

Although we are all aware of the link between cancer and nicotine, some tobacco myths, such as "smoking cigarettes is a stress reliever," are still accepted as gospel by smokers today. This myth was established by Professor Hans Selye, and it is illustrative of how the CTR worked.

Selye founded the International Institute for Stress in Canada. He published over 1,500 papers during his life (1907–1982), wrote 32 books, and was nominated for the Nobel Prize 17 times. Selye made stress research a legitimate scientific discipline. But Petticrew and Lee's research shows that in the last 20 years of his life, Selye was funded by the tobacco industry to throw doubt on legitimate science. He is responsible for introducing the incorrect belief that smoking relieves stress, at the behest of the tobacco companies.

In a 2017 interview about the negative effects of the CTR on stress research, Petticrew said, "I think what's often overlooked is the amount of money and effort the tobacco industry put into influencing popular culture. The tobacco industry threw so much money at so many key researchers in the field. You just cannot read that literature today without considering the extent to which the science of stress has been tainted by tobacco industry money."[4]

This strategy of producing scientific uncertainty in order to undercut public health efforts and regulatory interventions designed to reduce the harms of smoking damaged all science, not just stress-related research. The same tactics were adopted in the 1980s by Exxon and Shell in their battle against climate change research. Public trust in science and the scientific process was the collateral damage of these campaigns.

Undoing the damage done is no trivial undertaking. Scientists have to regain the public trust. We have to learn how to explain our findings in short, simple, and interesting (but still accurate) ways so that journalists and the media turn to scientists when they need comments. In turn, media courses have to teach aspiring journalists when opposing opinions are required and how to present science in an interesting way without having to resort to artificial confrontations. It is important that we all realize that although a political debate requires two opposing sides, a scientific consensus does not. When the media interpret "objectivity" to mean "equal time," those undermining scientific orthodoxy will win.[5] "Scientists must keep reminding society of the importance of the social mission of science—to provide the best information possible as the basis for public policy. And they should publicly affirm the intellectual virtues that they so effectively model: critical thinking, sustained inquiry and revision of beliefs on the basis of evidence."[6]

COMPLEXITY OF SCIENCE

The complexity and sheer volume of scientific knowledge are part of the reason the tobacco industry, climate change deniers, and others can undermine science. Our knowledge of science is rapidly increasing. This is great, because we know more and more about the workings of the world around us, which allows us to live longer, more energy efficient, and more comfortable lives. However, the same complexity and extent of scientific knowledge that make these advances possible have a number of drawbacks. People resent and fear the vastness of science, and legislators and regulators struggle with understanding the subtleties and keeping up with the pace required to maintain a competitive edge in science without endangering their constituents.

We have long passed the point where one person is capable of understanding all of science, and sentiments such as "only geniuses can have a career in science" and "as an adult, I don't see the point of now needing to understand science" are now commonly expressed.[7] This has resulted in a growth of science skepticism that has enabled the Trump administration and its followers to ignore scientists and scientific facts, particularly when talking about climate and environmental science.

Authors and thinkers such as Ray Kurzweil and Bruno Giussani think the problems are much larger. They suggest that science and technology are growing exponentially, while the structures of our society (government, education, economy, etc.) are designed for predictable linear increases and are dysfunctional in today's exponential growth.[8] This hasn't been much of a problem before because scientific knowledge was in the relatively flat initial phases of its growth, but now we have reached the steeply increasing section of the growth curve. Our nation-state model and our economic systems can't deal with the powers and dangers of artificial intelligence, gene drives, climate change, and other developments.

FAKE NEWS AND THE POST-TRUTH ERA

I am a professor of chemistry, have a PhD, and am active in scientific research, yet I frequently have to ask myself the question "Is this science or is it fiction?" I often struggle to answer it. Why is this so? The complexity of science is just part of the answer; the other part is the fact that we live in a post-truth era, a time when real news and fake news are sometimes indistinguishable; sometimes even when they are distinguishable, the facts are ignored. It has become so difficult to distinguish truth from fiction that some universities have felt the need to offer courses such as Living in a Post-truth World: How to Build Your Personal Baloney Detection Kit (University of California, Davis) and Bullshit Detection (University of Washington).

Fake news has been around as long as people have been spreading news, but social media gave it legs. The internet has changed the way science is communicated. In 1994, there were fewer than 3,000 websites. By 2014, there were more than 1 billion.[9] We have instant access to more news and information than ever before. Neighborhood gossip has been magnified by websites and social media. Facebook (2.1 billion users) and Twitter (300 million followers) have unprecedentedly large followings and have the potential to spread information further than any TV channel or newspaper. Science knowledge has simultaneously been democratized and devalued in this social media age. The line between news, knowledge, and entertainment has disappeared, creating the illusion of being informed.

In 2016, "post-truth" was the *Oxford English Dictionary*'s word of the year. According to the dictionary, it is an adjective defined as "relating to or denoting circumstances in which objective facts are less influential in shaping public opinion than appeals to emotion and personal belief." Today, as Kathleen Higgins, professor of philosophy at the University of Texas, Austin, writes, "Public tolerance of inaccurate and undefended allegations, non-sequiturs in response to hard questions and outright denials of facts is shockingly high."[10]

Fake news and post-truth are closely linked. A Google news search for the term "fake news" gets over 22 million hits. The CBS show *60 Minutes* defines fake news as "stories that are provably false, have enormous traction [popular appeal] in the culture, and are consumed by millions of people."[11] The *Collins Dictionary* reported that usage of the term "fake news" increased by 365 percent from 2016 to 2017. Because the term has been used as a political weapon and become so prevalent, we now have an alternative term for the phenomenon: false news.

Between false news, fake news, and post-truth, science is taking quite a beating. How did this come about?

Throughout the 1970s, Americans got their news from a limited number of sources. There were just three main TV networks (ABC, NBC, and CBS). Most national and international newspaper articles were sourced by the two major wire services, the Associated Press and United Press International. Consequently, all major news coverage was fairly similar and consistent. Furthermore, under Federal Communications Commission (FCC) fairness doctrine rules, broadcasters had to ensure that their coverage was fair and balanced.

Large changes occurred in the 1980s. Cable television (CNN debuted in 1980 and Fox News in 1986) and talk radio (the Rush Limbaugh show started broadcasting in 1986) started making their presence felt, and in 1987 the FCC eliminated its fairness doctrine. Radio and TV stations no longer had to present their news in a fair, factual, and honest manner.

Before the emergence of the internet and the elimination of the fairness doctrine, print and broadcast media adhered to FCC and self-imposed journalistic norms of objectivity and balance. The internet has enabled the rise of inexpensive alternatives to the established news sources. Many of these news outlets do not have editorial processes designed to establish the accuracy and credibility of the information they publish. They present "misinformation," which is incorrect information not recognized as false and spread inadvertently, and "disinformation," which is inaccurate information that is purposely spread to deceive people. These are the fake news outlets. Disinformation is spread to increase political influence, sell products and ideas, and deceive readers into clicking on links, maximizing traffic and thereby increasing the profit of the fake news outlets.

According to Aelive Digital Marketing, CNN is the top liberal website and Fox News is the most popular conservative website. Although both are fairly mainstream news outlets, they still have a high percentage of misrepresentation of facts. PolitiFact has found that 27 percent of CNN assertions were mostly or completely false, and more than half of Fox News' information was incorrect. Today risk and controversy sell news and generate advertising revenue.

Trolls and bots also contribute to the problem; they are the internet equivalent of the soapbox speakers who hawked their conspiracy theories on street corners. Trolls use social media to confuse complex news stories by adding incorrect facts, release inflammatory statements to start arguments, and strive to polarize debates. Many trolls are automated computer programs called bots, designed to spread disinformation and discord through social media networks. Facebook estimates that as many as 60 million of its users might be bots. Today, over 47 percent of Americans get their news from social media.[12] Not a reassuring statistic when we know that in the 2016 election around 20 percent of all tweets were generated by bots,[13] and that a 2018 study published in the *American Journal of Public Health* reported that Russian trolls and sophisticated bot accounts tweeted a plethora of vaccine-related messages. The aim of the tweets was not to influence the vaccine debate, as there were equal numbers of pro- and anti-vaccine tweets; instead they were designed to increase the polarization of American society by sending out tweets using divisive language linking vaccination to racial and social disparities.[14]

In March 2018, Soroush Vosoughi, Deb Roy, and Sinan Aral, all at MIT, published a paper titled "The Spread of True and False News Online" in *Science*.[15] They examined the spread of 126,000 true and false news stories that were tweeted and retweeted by approximately three million people. The veracity of the news stories was determined by six independent fact-checking organizations (including Snopes, PolitiFact, and FactCheck.org). The research showed that both scientists and nonscientists are more likely to repeat (in person or on social media) fake science news than real science news. It's not hard to see why; imagine you are looking up some information about a common neurological disease. You find 15 reputable-looking sites. They all have very similar information, except one has a new factoid you haven't seen on any of the other sites. This is the information you are likely to retweet or pass on to your friends. The fake news is more novel and often evokes more emotion than the real news. Even the most reliable and steady Twitter users occasionally succumb and retweet false news. These tweets often stand out from all the other tweets sent out in the prior 60 days. As Robinson Meyer writes in the *Atlantic*, "On platforms where every user is at once a reader, a writer, and a publisher, falsehoods are too seductive not to succeed: The thrill of novelty is too alluring, the titillation of disgust too difficult to transcend."[16] In a similar vein, long before the advent of the internet, Hannah Arendt wrote, "Lies are often much more plausible, more appealing to reason, than reality, since the liar has the great advantage of knowing beforehand what the audience wishes or expects to hear."[17]

On Twitter, news can be spread by one person with many followers tweeting the news item, which is then retweeted by a portion of the followers. It also can be started by someone with a much smaller following; the news gets read and retweeted, these retweets get retweeted, and so forth, producing much deeper penetration. Fake news beats real news in both the broad and deeper penetrations of the twittersphere. Other findings worth noting are that bots retweet true and false news at equal rates. Humans are much more susceptible to false news, and we are the ones who give fake news its legs. Accurate Twitter users have more followers than those who regularly tweet fake news. Finally, a breakdown of fake news into different categories reveals that most false tweets are political in nature, followed by urban legends, business, terrorism, science, entertainment, and natural disasters.[18]

To detect fake science, it is good to know why someone would want to distribute made-up scientific data. Unfortunately, I can think of numerous actors, aside from the previously mentioned tobacco lobby and other industries, who would benefit from spreading fake information, including the petroleum and coal industries, which are threatened by scientific results and try to undermine them with fake science. There are people and companies that use fake science to sell their products and people who use pseudoscience to argue for their irrational beliefs about vaccines, genetically modified foods, and homeopathy (chapter 11). Finally, there are those who use crazy, spectacular, and often false scientific results and facts to make us follow links to trashy stories that lead to further links. This is clickbait fake science, distributed by companies whose only income is derived from advertisers' fees when people click on links to stories containing their advertisements.

PREDATORY JOURNALS: BYPASSING PEER REVIEW

Chapter 4 mentioned predatory journals that publish papers without peer review, at a price, of course (typically $100–$2,000). German journalist Svea Eckert and her colleagues examined predatory journals and the conferences they organize. They found that the majority of the papers come from India, followed by the United States, including 162 papers from Stanford and 153 from Yale. Most of the papers come from academic institutions, but many are published by medical and biotech companies trying to hype their untested and unproven products.

In her autobiography *Dr. Hoffnung* (Dr. Hope), German TV host and actress Miriam Pielhau describes how she battled breast cancer in 2008 only to have it attack her liver in 2015. It was an aggressive growth, and the medical treatment she was getting was not successful. In desperation, Pielhau searched the scientific literature and came across a series of articles published about an anticancer drug produced by Immuno Biotech Ltd., called GcMAF. In the book she describes her newfound hope, an optimism based on her reading of scientific literature about GcMAF.

Pielhau died on July 12, 2016. The articles she had found, the ones that gave her hope, were published in predatory journals. No legitimate papers on GcMAF have been published in peer-reviewed journals, and it is not FDA approved. After she heard of Pielhau's case, Eckert wrote a fake article that claimed beeswax was a more effective treatment than conventionally used chemotherapies. It was accepted and printed by the *Journal of Integrative Oncology*. Having proven to herself that it is easy to publish dangerous misinformation, Eckert and her colleagues went on to document how the tobacco company Philip Morris, the pharmaceutical company AstraZeneca, and the nuclear safety company Framatone have used predatory journals and conferences to provide their research with a sheen of legitimacy gained from publishing in "peer-reviewed" science journals.

Alan Finkel, Australia's chief scientist, holds the opinion that "if journals are the gatekeepers of science, then predatory journals are the termites that eat the gates and make the community question the integrity of the structure." His suggestion to help readers, especially journalists, judge whether a paper has been published in a legitimate journal and not a predatory one is to create an internationally recognized stamp of recognition, similar to the fair-trade stamp of approval given to coffees: a signal that a publication is P-R (peer reviewed), not PR (public relations).[19]

TRUTH AND THE ROLE OF SCIENTISTS

After their own paper was misrepresented in the media, Chris Chambers and his colleagues at Cardiff University set out to establish whether the misreporting observed in the media could come from academia itself. Were university press releases exaggerating claims to enhance their universities' reputations or to attract venture capitalists to their start-ups? To answer that question, they examined all health-related science press releases issued by the top 20 universities in the United Kingdom in 2011, read the associated peer-reviewed articles, and then read the resultant news stories. They then tried to identify who was inserting the embellishments. Did the exaggerations originate in the newspaper articles, were they in the university press release, or were they added by the journalists covering the story? After reading and analyzing 462 papers and their associated 462 press releases and 668 news reports, Chambers and colleagues concluded that most of the exaggerations originated in the press releases, that the news reports carried forward the exaggerations, and most important, that there was no correlation between exaggerations and news coverage.[20] A typical exaggeration they encountered was that newspapers reported results as if they had been obtained through research done on humans when they actually came from research on model organisms such as mice. Another favorite was the age-old temptation to imply that correlation implies causation; for example, just because there might be a correlation between red wine consumption and length of life, that doesn't necessarily mean that something in red wine is responsible for the longer life. It could just be that red-wine drinkers are wealthier and have better health care. Time.com once even ran the headline, "Scientists Say Smelling Farts Might Prevent Cancer," based on a peer-reviewed article that merely pointed out that certain sulfide compounds are useful pharmacological tools to study mitochondrial dysfunction. It seems a straightforward solution would be closer cooperation between the researchers and the university communications departments when writing the press releases, in order to help make sure the news outlets won't latch onto misinformation.[21]

The increased access to scientific knowledge at the tips of our fingers has many advantages—it has certainly made writing this book much easier—but it has numerous disadvantages, too. Google searches make all of us mini experts. I suspect medical doctors have borne the brunt of this, as patients are continuously self-diagnosing and second-guessing their doctors. Scientists are also losing some of their authority to the web and Google. This wouldn't be a problem if all the science on the web was verified; unfortunately the good stuff (the peer-reviewed research) is hidden behind paywalls and not available to the casual searcher. The easily found material is rarely written by scientists. Some scientists respond to the challenge by dismissing and questioning the validity of alternate sources. This tends not to work. Some argue that scientists themselves need to step forward and present their research in an accessible manner. But research has shown that many scientists are not interested in educating the public and are contemptuous of their colleagues who try to do so. Comments such as the following, which was posted on a discussion forum for an online essay about making science accessible, are common: "I would love to explain (my research to the public) but I cannot. I cannot teach my pet hamster differential equations either."[22] Many of my colleagues are not quite as vocal and facetious as this, but they also believe it is not possible to simplify and make science accessible without losing important inherent subtleties. They fear that by popularizing science, we give the impression that science is easy, and readers mistakenly get the idea that they understand the science. I think this may be a small price to pay. At a 2016 workshop, "Social Media Effects on Scientific Controversies," Kevin Folta of the University of Florida pointed out that among researchers, "there is a disconnected arrogance that turns off the public and does not get them excited about learning more. Social media and the internet are a conduit of bad information. On social media it's easy to find information that scares you and scientists are not participating in trying to make it right."[23] Surveys have shown that nearly 40 percent of scientists vow never to use Twitter or Facebook for academic purposes, but this in an era in which the president has shown how effective Twitter can be to push his message.[24]

CONSEQUENCES AND SOLUTIONS

To see the effects of fake news, predatory publishers, and the systematic undermining of science by tobacco and energy companies, let's look at U.S. and global surveys that have been done to determine whether the general populace trusts science.

A survey of the state of science across 14 countries, commissioned by 3M, the adhesive company responsible for Post-It notes, found that 32 percent of the respondents were skeptical of science, and this group tended to drive negative and indifferent perceptions of science. Sixty percent of this group believed that their everyday lives would be no different if science didn't exist.[25] It's unclear where they think the technology to make things like their cell phones and TV screens came from and how there would be any medicines without science.

A 2016 Pew Research Center survey of adult Americans found that no more than a quarter of them had "a great deal of confidence" in scientists, and most had just "a fair amount of confidence" in them. Although this doesn't sound like overwhelming support, it is not that bad when one considers the fact that U.S. adults don't seem to be very trusting overall. Of the 12 professions evaluated, scientists came in second. The military came in first. Not surprisingly, I suppose, elected officials and business leaders came in last. More than half the people surveyed had a fair amount of trust in science and scientists; however, their belief was selective, especially when it came to issues related to climate change, childhood vaccines, and genetically modified food.[26] They were science deniers in certain areas where political, economic, and religious interests came into play. People without a high school degree are much less likely to have "a great deal of confidence in the scientific community" than those with an advanced degree (28 versus 61 percent). There is also a political divide. In 1974, 56 percent of conservatives had a great deal of confidence in scientists; that number had dropped to 36 percent by 2016. Confidence in the scientific community among liberals remained constant over the same period.

How do we deal with the fake science problem? Most important, we have to learn to recognize it and not fall for its novelty. If the science sounds too good to be true, is too wacky to be real, or very conveniently supports a contentious cause, then you might want to go to the source of the research and see whether it was published in a peer-reviewed journal and what the original paper actually says. However, just recognizing fake science treats the symptoms rather than the problem. We need to combat fake science through concerted efforts by scientists, journalists, social media companies, and readers: the entire complex web.

Scientists have to do more outreach; they need to be more accessible to the general public and work more closely with those writing press releases.

The current system is an untenable version of the telephone game, in which a very complicated message originates from a scientist, who uses a lot of jargon to assure accuracy, then goes through the public relations department to journalists, news copy editors, and finally readers. Scientists have a strong vested interest in combating fake science. We need public support that will ultimately determine the amount of federal funding allocated to research; we also need public trust, and as scientists we need to be able to trust the literature we read. If you have ever worked on a puzzle and put a piece in the wrong place, you know it totally messes up the remainder of the puzzle—that is what happens if researchers build their projects on fake science.

By definition, journalists hunt stories and critically analyze facts; in the future they have to extend their skepticism to scientific press releases. This is increasingly difficult, as there are fewer and fewer science journalists employed in the newspaper business. In addition, news copy editors, who write most headlines, must resist producing misleading headlines.

Facebook and Twitter need to limit the number of bots on their sites and institute rules that control trolls. Artificial intelligence programs should be used to find fake news (and fake science) so that it can be removed from the sites. Facebook, Twitter, and Google are trying to limit the amount of fake news on their sites, but they are playing their cards close to their vests, and we don't know what they are doing. For instance, Google was considering a "truth rating" to accompany and guide Google searches. This idea was dropped after conservative lobbyists convinced the company that a truth rating would be discriminatory to their causes.

In May 2019, Singapore passed the Protection from Online Falsehoods and Manipulation bill, which bans the spreading of "a false statement of fact" harmful to the public interest and allows the appropriate authorities to order social media platforms to remove offending posts and issue apologies. Violations can result in big fines (~$750,000) and up to 10 years' imprisonment. Free speech nongovernmental organizations (NGOs) all around the world have condemned the law, as it could be used to clamp down on peaceful antigovernment protesters and will result in the government deciding what is true and what is not. To combat such draconian laws, we as readers have to be more critical.[27]

We need to resist clicking on clickbait. Each time we follow a clickbait link, the purveyors of fake science make money.

Predatory publishing is a very profitable business. Tenure and promotion committees, funding sources, journalists, and readers need to be very leery of any papers published in predatory journals. Predatory publishing will disappear when the only people who publish in predatory journals are hucksters trying to legitimize pseudoscientific ideas. To get to this point, we have to change the way research is evaluated; if we stay with a model that heavily relies on the number of papers published, we need a curated, accepted list of legitimate peer-reviewed journals.

New scientific breakthroughs made in the fake news era will increase the quality of our lives and that of our children. However, these techniques will also allow scientists to do experiments that border on the fantastic, further increasing the difficulty in distinguishing between fact and hyperbole.

Chapter Eleven

This Is Science, Not Politics

You idiots. Grow the fuck up you're not children anymore. I didn't mind explaining photosynthesis to you when you were 12. But you're adults now, and this is an actual crisis; got it?—Bill Nye, talking about climate change

Sometimes people have very strong beliefs, identities, or biases. They can be strong enough to overcome rational analysis in even the most sophisticated scientists. Despite having all the facts, training, background knowledge, and scientific competence, both scientists and nonscientists took a long time to acknowledge that the earth was round, that it rotated around the sun, that species evolved over time, and that women scientists are just as capable as men (and even these statements remain more contentious among the public than they should be). Charles Darwin's contemporaries had the scientific chops to understand his theory of evolution, but their religious beliefs were so ingrained that they were not able to accept it. Climate change, the efficacy of vaccinations, and the safety of genetically modified (GM) foods are today's flat earth issues (although trolls would have us believe that the flat earth theory is still alive and well). A substantial number of people in the world do not believe the science associated with climate change, vaccinations, and GM foods because of their political and cultural identities. In this time of false news and fake science, it is easy and convenient not to believe proven science. This is a problem from the right (climate change) all the way to the left (GM foods). The consequences of real science meeting fake politics can be deadly.

CLIMATE CHANGE

In 1827, French mathematician Joseph Fourier proposed a radically new theory, the greenhouse effect. He suggested that gases in the atmosphere naturally trap heat on Earth in the same way that a pane of glass in a greenhouse traps heat.[1] Other research has since proven that he was right. Without atmospheric gases trapping the infrared radiation in Earth's atmosphere, the average temperature of Earth would be −18°C (−0.4°F). All water would be frozen, and there would be no life as we know it. Scientists have proven that the greenhouse effect warms Earth to an average of 15°C (59°F), which allows a variety of life-forms to survive. Fourier was also one of the first scientists to point out that human activity will change the amount of thermal radiation (heat) trapped in the atmosphere, leading to climate change.

About 60 years after Fourier proposed the greenhouse effect, Swedish chemist Svante Arrhenius published an article titled "The Influence of Carbonic Acid (Carbon Dioxide) in the Air upon the Temperature of the Ground,"[2] in which he described how carbon dioxide in the atmosphere traps heat. He also accurately calculated the temperature increases caused over time by the rise in carbon dioxide in the atmosphere. From these calculations, he concluded that Earth's glacial periods, when large parts of the planet were covered with thick sheets of ice, were caused by a reduction of the amount of carbon dioxide in the air. Today's scientists agree, and carbon dioxide and other greenhouse gases in Earth's atmosphere are increasing in concentration as a result of human activities. As a consequence, the average temperature of Earth is increasing beyond 15°C, and this is leading to climate change.[3]

Wallace Broecker published an article in 1975 in *Science* titled "Climatic Change: Are We on the Brink of a Pronounced Global Warming?," which introduced both terms that have been used to describe this change in climate. "Global warming" was most used at first. But in the 1980s, the United States and Saudi Arabia successfully lobbied the world climate negotiation groups for use of the term "climate change," which they felt was less threatening and had fewer connections to the burning of fossil fuels.[4]

Carbon dioxide and other gases that absorb the infrared radiation emanating from the sun and the Earth are called "greenhouse gases." Climate change is caused by the increasing concentration of these gases in the atmosphere, which then trap more heat. Greenhouse gases concentrate in the atmosphere when the sources and amounts of gas emissions are greater than their sinks. A "sink" is a process or organism that removes a greenhouse gas from the atmosphere. When the source and the sink are about equal, the concentration of a greenhouse gas will remain at equilibrium, or constant.

The three main sources of carbon dioxide in Earth's atmosphere are the burning of fossil fuels (combustion); trees and other living organisms naturally exhaling carbon dioxide (respiration); and concrete manufacturing, in which limestone (CaCO$_3$) is heated to 10,000°C (18,000°F) to produce lime (CaO) and carbon dioxide. The lime is a critical component of cement. Large amounts of carbon dioxide are released as a by-product of the process and in the combustion required to reach 10,000°C.

Combustion of fossil fuels and concrete production have both dramatically increased since the Industrial Revolution. Earth also has more people (more than seven billion) than ever before. Those billions of people have millions of cars, thousands of planes, and innumerable power plants and factories that all rely on burning fossil fuels. With growing populations and cities, humans are also constructing more and more buildings that require concrete.

Carbon dioxide has two natural sinks: green plants and natural bodies of water. Trees, plants, and forests use carbon dioxide as a key part of photosynthesis, and the bodies of water naturally dissolve carbon dioxide. This means deforestation (which is typically used to make way for cattle farms, rice paddies, or human habitation) is removing a very large and important carbon dioxide sink while simultaneously replacing it with more carbon dioxide sources. The result of this cycle is simple: the concentration of carbon dioxide is steadily increasing.

Scientists measure the concentration of carbon dioxide in the atmosphere in parts per million (ppm), or the number of CO$_2$ molecules in every million molecules of gas. In preindustrial times, the average world CO$_2$ levels were about 280 ppm. In 2016, the levels passed 400 ppm for the first time in four million years. That means that for every 1 million gas molecules in the atmosphere, at least 400 are CO$_2$ molecules.

Most of the world's population lives in the Northern Hemisphere, so the greatest volume of carbon dioxide emissions come from this part of the planet. Carbon dioxide emissions stay in the atmosphere for years because it takes time for the gas to dissolve in seawater and for trees and plants to absorb it. Meanwhile, air currents spread the CO$_2$ all over the world, so carbon emissions are a global problem. "The increase of carbon dioxide is everywhere. . . . If you emit carbon dioxide in New York, some fraction of it will be in the South Pole next year,"[5] says Pieter Tans, a senior scientist at the National Oceanic and Atmospheric Administration's (NOAA) Earth System Research Laboratory in Boulder, Colorado. Even the carbon dioxide levels at all the South Pole research stations have surpassed 400 ppm.

Carbon dioxide is a colorless, odorless, and nontoxic gas. It is therefore difficult to visualize and easy to ignore: not ideal properties for a great villain. It is hard to drum up support for the regulation of this by-product of combustion; a smelly, acrid gas would have been much easier to regulate.

Methane is another greenhouse gas that impacts Earth. It is actually 50 times more effective at absorbing infrared radiation than the same mass of carbon dioxide. Yet methane currently causes only one-third the amount of global warming that carbon dioxide does because humans and animals produce less of it.

Methane comes from a variety of sources. If organic matter decays in the absence of oxygen, it produces methane. For example, methane (sometimes called marsh gas) is produced when plant matter decays in the stagnant (deprived of oxygen) waters of a wetland or marsh. In fact, one-quarter of all methane emissions in the atmosphere naturally bubbles up from wetlands. The other major sources of methane are anthropogenic (created by humans). They are associated with farming (cows and rice paddies). Crude oil and natural gas contain methane, which easily escapes into the atmosphere. Methane is released from the time fossil fuels are taken out of the ground until they are finally used. It is released during mining, refining, transport, fracking, and energy production.

Livestock animals are a major nonhuman source of methane emissions, contributing about one-third of total methane emissions. Many animals, particularly cattle and sheep, have a rumen (the first part of their stomach) in which bacteria and other microorganisms break down cellulose from the grasses and leaves they eat. A by-product of this digestive process is methane, which the animals release through belching and flatulence. An average cow emits 250 liters (66 gallons) of methane every day, and Earth has about 1.5 billion cows. That's about 375 billion liters (100 billion gallons) every day![6]

Rice paddies are much like wetlands. Rice requires a lot of water to grow, but most of the water in the paddies is stagnant. With little oxygen in the standing water, small plants or weeds in the paddies decay and release methane.

The only sink for atmospheric methane is the hydroxyl radical. This radical is formed by ultraviolet (UV) light reacting with oxygen and water. The reaction lasts less than a second, and its concentration has been fairly constant over time. So, while methane emissions are increasing, the sink is remaining constant. This results in a net increase in methane.[7]

The National Aeronautics and Space Administration (NASA) does a monthly analysis of global temperatures.[8] Researchers assemble the data from various sources. They include 6,300 meteorological stations around the world, ship- and buoy-based instruments measuring sea surface temperatures, and Antarctic research stations. Sixteen of the seventeen warmest years in the 136 years of NASA temperature records have occurred since 2001. (The one exception was 1998.) In fact, the year 2016 had the hottest global average temperature—0.94°C higher than the 1880–2017 average temperature—in recorded history. This high temperature was largely due to increases in carbon dioxide and methane levels. A strong El Niño—a warming phase of a Pacific Ocean climate cycle—added to the heat that year.

The average global temperature on Earth has increased by about 0.85°C since 1880. That was the year the International Meteorological Organization began standardizing the recording of global temperatures. Two-thirds of that warming has occurred since 1975, at a rate of roughly 0.15°C–0.20°C per decade.

On June 23, 1988, Dr. James Hansen, then director of NASA's Institute for Space Studies, testified before the U.S. Senate Energy and Natural Resources Committee: "Global warming has reached a level such that we can ascribe with a high degree of confidence a cause-and-effect relationship between the greenhouse effect and observed warming. . . . In my opinion, the greenhouse effect has been detected, and it is changing our climate now."[9]

The North and South Poles have experienced the largest temperature increases. The warming has led to significant melting of sea ice. Perennial, or year-round, sea ice in the Arctic is declining at a rate of 9 percent per decade. Long-term temperature forecasts predict a rise of 1.4 to 5.6 degrees Celsius (2.5 to 10 Fahrenheit) over the next century.[10] This may seem like a small temperature change, but a 1- to 2-degree drop was all it took to move the earth into the Little Ice Age 700 years ago. As predicted in earlier studies, these small changes can have significant local changes. As temperatures warm, glaciers are shrinking, ice on rivers and lakes is breaking up earlier every year, plant and animal ranges are shifting, and trees are flowering sooner each year. The melting of sea ice results in an accelerated global rise in sea levels. Heat waves are longer and more intense. Scientists have shown that climate change is associated with an increase in extreme weather events. Most, but not all, of these events are due to temperature increases. For example, NOAA has recorded the number and the cost of all weather-related disasters that have occurred in the United States since 1980. From 1980 to 2017, NOAA recorded 212 weather disasters that have cost more than $1 billion each. The total cost of the events exceeded $1.2 trillion. The 1980–2016 annual average was 5.5 events per year. Due to climate change, it increased to 10.6 events per year for the 2012–2016 period.

Those are the facts. It is a commonly held belief that the more science knowledge one has, the more one supports science. This belief has been borne out by a number of studies, and some Pew Research Center surveys conducted prior to 2016 also lend support to that theory. However, a 2016 opinion poll found that this theory is not always valid. When Republicans were asked whether climate change was due to human activity, 19 percent of those with low science knowledge, 25 percent with medium science background, and only 23 percent with high science knowledge agreed there was a link. Well-educated Republicans with an extensive science background were just as likely to have no confidence in climate science as Republicans without science knowledge. It seems our scientific beliefs in some areas have been politicized.

Democrats with an average better scientific understanding are more likely to acknowledge that climate change is linked to human activity than those with a lower understanding of scientific issues; the numbers range from 49 percent for those with low scientific understanding to 93 percent for those with a high science background.[11] The poll also found that people who deeply care about climate change are more likely to know that carbon dioxide is produced as a consequence of burning fossil fuels (75 vs. 65 percent).

A 2018 Gallup poll confirmed the Pew survey results showing that there is a party divide. They found that 91 percent of Democrats say they worry a great deal or a fair amount about climate change, while only 33 percent of Republicans say the same.[12]

Climate change denial is an identity. It is a sign that one is a conservative and is a social cue that shows membership in that group. Liberals have their own identifiers, such as a distrust of genetically modified foods. We like to be in our group. A clever experiment in Britain asked subjects to compare themselves with people in Sweden and then asked them about their thoughts on energy conservation. It compared their results to a similar group who had measured themselves against people from the United States. The group that had been matched with the Swedes, who are known to be very environmentally aware, was less interested in being energy conscientious than the group comparing themselves with the Americans, who are perceived as energy wasters. The Brits unconsciously moved away from the comparison group to stay in their own group.

Dan M. Kahan is the Elizabeth K. Dollard Professor of Law at Yale University and is known for his theory of cultural cognition. He believes that climate change science has become contaminated with group identity and social meaning. He compares it to gun control in West Virginia, where 65 percent of the people want more gun control but would never vote for someone who supports gun control, because "85 percent of the people in West Virginia know that you can't trust politicians who say they want gun control."[13]

A previous Pew poll taken in 2014 compared the opinions of a random sample of the general public with a representative sample of scientists connected to the American Association for the Advancement of Science (AAAS). One of the largest disconnects between scientists and nonscientists was recorded when they were asked whether they believed that "climate change is mostly due to human activity." There was a 37 percent gap between scientists, 87 percent of whom answered "yes," and nonscientists, only 50 percent of whom thought that climate change was due to human activity.[14] The gap is even larger if one considers just climate scientists, 97 percent of whom agree that global warming is caused by humans.

Belief in human-caused climate is fickle. A University of New Hampshire survey of political independents found that 70 percent believed that human activity caused climate change on a very hot day, but that number fell to 40 percent on very cold days.[15] Weather forecasters are on the front lines of climate research communications, but most of them are not trained climatologists, and it is interesting to see what they think. Surveys of weather forecasters show a promising increase in the belief that anthropogenic climate change is occurring: the numbers rose from 22 percent in 2001, to 52 percent in 2010, and to 80 percent in 2017.[16]

Climate skepticism is mainly limited to English-speaking countries. According to James Painter, a research associate at the Reuters Institute, "We looked at a very large number of articles, more than 3,000, and more than 80 percent of the articles that had climate skepticism in them were found in the US and the UK compared with the newspapers in Brazil, China, India and France."[17] Most of the climate change denial occurred in more right-leaning newspapers and television stations. However, most English-speaking media outlets are guilty of "false balance," in which fringe views are given equal time with legitimate climate scientists. In China, the media and government move in lockstep, and since the Chinese government has long accepted the link between global warming and human activity, the media represent climate scientists fairly. It must be noted that although China is the largest producer of carbon dioxide in the world, it is not the largest producer of carbon dioxide per person in the world; that honor belongs to the United States. Hans Rosling, one of the authors of *Factfulness*, thought comparing total carbon dioxide production by nation was useless, a bit like "claiming that obesity is worse in China than in the United States because the total bodyweight of the Chinese population is higher than that of the US population. Arguing about emissions per nation is pointless when there is such enormous variation in population size."[18] *ɪᴛ ᴀʟsᴏ ᴅᴏᴇsɴ'ᴛ sᴏʟᴠᴇ ᴛʜᴇ ᴘʀᴏʙʟᴇᴍ*

Thirty years ago, when Hansen testified that human activity was responsible for climate change, the impact was dramatic. Newspapers, TV stations, politicians, corporations, and nongovernmental organizations (NGOs) recognized the importance of the problem, and climate change was discussed in a largely nonpartisan fashion.[19] Since his announcement, scientific certainty that human causes are responsible for climate change has become stronger. As predicted, we have seen more wildfires in the western United States, more flooding associated with hurricanes in the southeastern United States, rising sea levels, bleaching of corals, spreading of tropical diseases, and increased ocean acidification, yet nothing has been done in the United States. There have been over 600 congressional hearings on climate change, but little progress has been made, especially after President Donald Trump withdrew from the international Paris Agreement on Climate Change. At the same time, concern about climate change has decreased among Republicans. Why? Because in 1989 the petroleum industry took a page out of the tobacco industry playbook (chapter 10) and started the Global Climate Coalition, which began a well-financed campaign to undermine climate science research by polarizing public opinion, creating misinformation, and promoting uncertainty. The 1992 Earth Summit occurred a year after the collapse of the Soviet Union, and fearmongers had a ready replacement for the "red menace." Rush Limbaugh proclaimed that climate science "has become a home for displaced socialists and communists," and the Heartland Institute in Chicago erected a billboard depicting Unabomber Ted Kaczynski with the caption, "I still believe in global warming, do you?"[20] By default, climate scientists had become the enemy through these demonizing tactics. It was okay to shoot the messenger. The coordinated efforts of conservative foundations and the petroleum industry officially ended in 2001. However, the nonprofit Union of Concerned Scientists has revealed that ExxonMobil continued to fund conservative think tanks that were willing to "ensure that the recognition of uncertainties of climate science becomes part of the 'conventional wisdom.'"[21]

Even in countries where climate change hasn't been used as a tool to "stir up the base," it is difficult to regulate the release of greenhouse gases, for a number of reasons:

- It is much easier to prevent the start of something new than it is stop something as entrenched as our energy consumption. At least once a week I drive by a nuclear power plant that is located a few miles from our house. I hardly register that it is there; occasionally I think of Fukushima and Chernobyl and the potassium iodide pills we have. No one protests its existence, but there is no way anyone could build a new nuclear power plant in our neighborhood. Cars, planes, meat consumption, and electricity use are so interwoven into our lives that it is difficult to change our dependence on them.
- Climate change is a nebulous enemy; it is complex and multifaceted and doesn't make for good TV, especially if we remove the false balance. Mark Brayne, a former BBC senior correspondent, says a good story requires "a narrative of baddies and goodies," but he complains that climate change has neither of these. "It is slow moving, complex and what's more, we ourselves are the baddies. That's not something listeners and viewers want or wanted to be told."[22]
- We overreact when exposed to very visible threats. Terrorism, carjackings, kidnappings, and mass shootings all invoke a sense of immediate alarm, and we respond with a loss of proportion; we overcompensate and don't consider the low probability of their occurring to us. But extreme weather doesn't have the same stigma; there is some familiarity. This gives us leeway to believe that it is just nature doing its thing. We can't see or even imagine the culprit: more than 400 colorless gas molecules dispersed among each million nitrogen and oxygen molecules.

- Not all countries will be affected by climate change in the same way. In fact, it is the smaller, poorer, and less influential countries that will be affected the most, especially small islands and low-lying countries like Bangladesh. The main carbon producers, the United States and China, are less vulnerable. Some models even have Russia coming out ahead with climate change. The Caribbean islands and Greenland have much greater incentive but much less power to negotiate a climate treaty than the United States does.

Naomi Oreskes, professor of the history of science at Harvard University, has some suggestions, discussed in "How to Break the Climate Deadlock."[23] She thinks that the United States has to recognize that the free market alone can't address climate change; we have to accept the fact that tackling climate change does not infringe on our liberties and concede that the markets need input from the government. In her opinion, we require "a different vision, one that embraces priorities other than profit, and places care—for creation and for one another—at its center." Bill McKibben agrees that the free market is not the solution; it has performed brilliantly at creating wealth, but it forgot that there are other tasks. He argues that economists are unlikely to embrace the care required to change our CO_2-producing lives.[24] He bases his opinions partly on research showing that economics students rate helpfulness, honesty, and loyalty as less important as they progress through their university careers; economics professors "give significantly less money to charity than their worse-paid colleagues in many other disciplines"; and "after taking a course in economic game theory, college students behaved more selfishly and expected other[s] to do so as well."[25]

In the 19th century, scientists and nonscientists were on the same page. Farmers, villagers, and climatologists were all concerned about the effects that increased agriculture and its use of water would have on future rainfall. Weather was widely discussed in bars, village squares, newspapers, the legislature, and scientific literature. Climatology was in its infancy and relied heavily on the general public, particularly farmers and sailors, for its weather information. This was an early version of citizen science. Austrian meteorologist Karl Kreil (1798–1862) was the founder of the largest climatological observation network of the 19th century. He proudly proclaimed his dual identity as scholar and public servant. Climatology was messy and multifarious. Deborah Coen, professor of history at Yale, argues that modern climate science has evolved (or perhaps devolved) from a hurly-burly, inclusive free-for-all to a monolithic field of mostly white male scientists and politicians. Consensus rules, both in politics and in science, where matching climate models are desired. In the face of skepticism, scientists and the Intergovernmental Panel on Climate Change have prioritized consensus building by reducing their statements to the vaguest, blandest terms. Coen says, "The consensus-building approach of recent climate science has successfully established anthropogenic climate change as an indisputable fact. But it has failed to translate that knowledge into action. The solution may lie in a return to pluralism."[26]

I have an acquaintance who does not believe that humans are responsible for the current changes in climate. He is very technosavvy; when I want a new phone or have a question about Alexa, he is the one I ask. We avoid talking about politics, but somehow, and I am never sure how, climate change does come up. In the beginning of the discussion, he will always have perfectly reasonable questions, for example, "I understand the science, but this could just be a naturally occurring warming trend, or what about aerosols and clouds formed by increased evaporation?" I always fall for it; knowing he is intelligent and scientifically knowledgeable, I give him the facts and explain why we know this is not a natural trend, and so forth. As his eyes glaze over, I remember that his identity as a conservative is stronger than his willingness to understand. I grew up in apartheid South Africa and remember seeing a similar reaction when my friends and I argued about politics.

Katharine Hayhoe is a professor and director of the Climate Science Center at Texas Tech University, Lubbock. She has also come across many climate change skeptics who have used sciency objections as a smokescreen to hide the fact that the real reason they question climate science has more to do with their identity and ideology than with data and facts. She has tried many ways of communicating with skeptics and has found that "the most effective thing I've done is to let people know that I am a Christian. Why? Because it's essential to connect the impacts of a changing climate directly to what's already meaningful in one's life, and for many people, faith is central to who they are."[27] Hayhoe also feels that to change minds, she not only needs to find a shared identity but also needs to offer hope: "Changing minds also requires providing practical, viable, and attractive solutions that someone can get excited about."[28]

You might have noticed that two of my openings begin with quotes by a comedian. That is no accident; not only am I easily amused, but I also agree with Robert B. Crease, who thinks that "comedians have an ability to speak truths—truths that people are afraid to talk about, that people can't even see—in a way that breaks through resistance, cuts through codes, and speaks truth to power. They have a license to be inappropriate."[29]

Kate Marvel is a climate scientist at Columbia University and NASA's Goddard Institute for Space Studies. She refuses to debate science in public, especially climate science. In her opinion, "Once you put the facts about the world up for debate, you've already lost. Science isn't a popularity contest; if it were, I'd definitely vote to eliminate quantum mechanics, set π to 1, and put radium back in toothpaste."[30] Climate change and the science behind it are fact, not opinion; this is not something we can "agree to disagree" about.

In the 2002 U.S. presidential elections, Republican candidates were briefed by communications specialist Frank Luntz. He said, "A compelling story, even if factually inaccurate, can be more emotionally compelling than a dry recitation of the truth."[31] And he advised the candidates to make the lack of scientific certainty about climate change a primary issue in the debates. He told them, "Should the public come to believe that the scientific issues are settled, their views about global warming will change accordingly. Therefore, you need to continue to make the lack of scientific certainty a primary issue in the debate."[32] Republicans heeded his advice and emphasized the inherent uncertainties in climate science, even though even larger uncertainties did not bother them in other situations. For example, prominent climate change skeptic Mitt Romney justified increased military spending by saying, "We don't know what the world is going to throw at us down the road. So we have to make decisions based upon uncertainty." And Dick Cheney, a virulent opponent of attempts to reduce CO_2 production, has said, "Even if there is only a one percent chance of terrorists getting weapons of mass destruction, we must act as if it is a certainty."[33] In contrast, the fact that no scientist could with any certainty say that a single weather event was caused by climate change has effectively shut down many a debate and attempt at passing related legislation.

In my opinion, attribution science might be the game changer we are all hoping for. After nearly a decade of research and more than 180 research papers, this field has matured enough that it is going public. In some cases, meteorological services will be able to tell us how much more likely an extreme weather event is due to the greenhouse gasses released since the onset of the Industrial Revolution. *Nature* reviewed all the climate attribution studies from 2004 to mid-2018, finding that two-thirds of the climate events examined were more severe due to the release of greenhouse gas related to human activities. Forty-three percent of these events were heat events, 18 percent were droughts, and 17 percent were extreme rain or flooding. Attribution studies showed that the 2016 heat waves in Asia, the global heat record in 2015, and the increased ocean temperatures in the Gulf of Alaska and the Bering Sea in 2014–2016 would not have occurred without anthropogenic climate change.[34] Friederike Otto, a climate modeler at the University of Oxford, uses weather@home, a citizen science distributed computing framework similar to folding@home (chapter 3), to do her attribution calculations. She is collaborating with the German national weather agency. According to Paul Becker, vice president of the weather agency, they want "to quantify the influence of climate change on any atmospheric conditions that might bring extreme weather to Germany or central Europe. The science is ripe to start doing it."[35] Quickly establishing a link between a climate event and climate change, or ruling it out, will be very effective. Otto says, "If we scientists don't say anything, other people will answer that question not based on scientific evidence, but on whatever their agenda is. So, if we want science to be part of the discussion that is happening, we need to say something fast."[36]

Life as we are living it is unsustainable. Our economic system, which is based on growth, has served us well, but perhaps its time has come. Maybe we have reached maturity, and it is time to stop growing. In *Fracture*, Bill McKibben writes:

> What worked in the past doesn't automatically work in the future. At one point, growth provided more benefit than cost. Light regulation spurred expansion. Larger scale offered efficiencies that made us richer. Fine. You want your child to grow—if she doesn't, you take her to the doctor. But if she's twenty-two and still shooting up by six inches a year, you take her to the doctor, too. There's a time and a place for growth, and a time and a place for maturity, for balance, for scale. . . . Perhaps our job, at this particular point in time, is to slow things down, just as basketball teams do when they're ahead. . . . Given that there is no finishing line to the human game, no obvious goal toward which we are racing, then why exactly are we so intent on constantly speeding up?[37]

Our current lifestyles, predicated on expanding economies, are generating too much waste. Plastic bags, flip flops, and plastic straws have been found in the deepest ocean trenches and the stomachs of whales, dolphins, and turtles. The photos and videos of plastic pollution go viral and evoke a visceral feeling of despair and disgust. That is why I am pretty sure plastic waste has reached its peak. It will be much more difficult to reduce the amount of carbon dioxide and methane released into the atmosphere. We can't see these gases; we have to believe scientists that their concentrations are increasing and that this will result in an increasing number of extreme weather events. Our belief in the science has to be solid and unwavering. Based on our belief in science, we are asking people to change the way they live their lives and their worldviews. To significantly change our greenhouse gas emissions, we all have to change where we get our electricity, how we travel, what we eat, and what we consume. This is a big ask; to save the world we have to trust the science and change our lifestyles. The only way this can be done is by making private, public, local, and global changes. Maintaining and bolstering trust in science has never been more critical.

ANTI-VAXXERS

The anti-vaccine debate is very similar to the one about climate change. In both cases the science is conclusive, and the denialists are strongly influenced by their group identity. Fortunately, anti-vaxxers are not as common as climate change denialists, and neither political party in the United States actively opposes vaccination. Anti-vaxxers tend to be on both ends of the political spectrum and are often religious groups, greens, or libertarians, who object to the fact that the state has the right to restrict individual liberty in order to promote welfare for all. They distrust pharmaceutical companies and think money corrupts medicine.

As long as there have been vaccines, there have been anti-vaxxers. In England, in his 1772 sermon, "The Dangerous and Sinful Practice of Inoculation," Reverend Edmund Massey called vaccines "diabolical operations."[38] Since then, the opposition to vaccines has never disappeared.

The acceptance of vaccines is very important, because vaccination rates of 95–99 percent are required to preserve herd immunity against highly contagious diseases like measles and chicken pox. In this way, even people who can't be immunized (e.g., young babies) are protected. Vaccines have helped decrease the prevalence of childhood diseases and have eliminated or nearly eliminated smallpox and polio. However, mistrust of vaccines may undo much of this progress, resulting in outbreaks of disease we thought we had defeated and endangering children who have medical conditions that prevent them from being vaccinated.

In February 1998, *The Lancet*, a high-impact, well-respected medical journal, published an article that reported a connection between the measles, mumps, and rubella (MMR) vaccine and autism in children. The work was done and reported by Andrew Wakefield, a gastroenterologist at the Royal Free Hospital in London. At the time not much was known about the causes of autism, and the Wakefield paper caused a sensation; it "went viral." Many parents of autistic children came forward to confirm that they first noticed the symptoms of autism in their children shortly after they had been inoculated with the MMR vaccine. The medical community was less impressed because the study involved only eight children, and it is well known that children, vaccinated or not, exhibit the first characteristics of autism at about the age they would normally get the MMR shot. The government tasked the British Medical Research Council to find out if there was any validity to the findings. The council found none, and a number of other papers have been published refuting the findings. Wakefield vocally and very publicly defends his conclusions and claims there is a conspiracy to silence him. Journalists covering the controversy found that his continued research was being funded by a group of lawyers, the Legal Aid Board, who were suing the manufacturers of the MMR vaccine. Wakefield hadn't disclosed his funding source to anyone, including his colleagues. Ultimately Wakefield was struck off the medical register and forbidden from practicing medicine in Britain, and *The Lancet* retracted his paper, saying it was "utterly false."[39] But the damage was done. The media grabbed hold of the story, and his ideas attracted many followers, including a number of high-profile Hollywood stars. Actress Jenny McCarthy was particularly taken by Wakefield's story. She has an autistic son, Evan, and is convinced it was the MMR shot that "gave" him autism. In 2007, she went on the *Oprah Winfrey Show* and told America about the "dangers" of vaccines. In the days after the show she also appeared on *Larry King Live* and *Good Morning America*. In *The Panic Virus*, Seth Mnookin estimates that McCarthy's anti-vaccine message reached 15–20 million people through those shows.[40] In 2011, McCarthy wrote the foreword to Wakefield's book, *Callous Disregard: Autism and Vaccines—The Truth Behind a Tragedy*.

After being disgraced in the United Kingdom, Wakefield moved to Austin, Texas, and founded the Autism Media Channel, which makes videos asserting a link between autism and the MMR vaccine. He has also turned to social media to spread his message. "In this country, it's become so polarised now. . . . No one knows quite what to believe," Wakefield said in an interview with *The Guardian*. "So, people are turning increasingly to social media."[41]

Vaxxed, a film directed by Wakefield, was scheduled to premiere at the Tribeca Film Festival. It briefly enjoyed strong support from Robert De Niro, who also has an autistic child. The film describes the alleged cover-up of the link between MMR and autism by the CDC. After a media uproar and meeting with scientists, De Niro withdrew the film from the festival. Despite being discredited by the medical and scientific communities, Wakefield still has a strong celebrity following; he spoke at one of Trump's inaugural balls, and in 2018 he was dating Australian supermodel Elle Macpherson, who has her own line of wellness products.[42]

Not only are Wakefield and the anti-vaxxer celebrities very good at using social media, but other players are using the discord created by the anti-vaccination debate to their advantage. As mentioned in chapter 10, a 2018 study in the *American Journal of Public Health* analyzed vaccine-related tweets between July 2014 and September 2017 and showed that Russian bots and trolls seemed to be quite active in sending both anti-vaccine and pro-vaccine messages on Twitter (to increase the volume of highly polarizing statements).[43]

According to the WHO, a drop in measles (MMR) vaccinations in Europe, attributed to the Wakefield paper and anti-vaxxer movement, resulted in a fourfold increase in measles in 2017. "Over 20,000 cases of measles, and 35 lives lost in 2017 alone, are a tragedy we simply cannot accept," said Dr. Zsuzsanna Jakab, the WHO regional director for Europe, at the time.[44]

All 50 states in the United States require students to be vaccinated before they enter school. Every state also allows medical exemptions for sick children who can't be vaccinated; they are protected from contagious diseases by the herd effect. The fact that all their classmates are vaccinated means there will be no disease outbreaks and no danger for the more vulnerable children to catch a preventable disease. Twenty states allow parents to forgo vaccinating their children for personal or philosophical reasons, and most states have religious exemptions. According to the CDC, the number of children under age two who have not been vaccinated has quadrupled since 2001. This is so despite the fact that research has shown unvaccinated children are 23 times more likely to develop whooping cough and 9 times more likely to be infected with measles and that their incidence of hospitalization is 6.5 times higher than that of vaccinated children from the same communities.[45] The CDC has also estimated that since 1998, vaccinations have prevented over 20 million hospitalizations and 730,000 deaths. In *Factfulness*, global health expert Hans Rosling stated, "In a devastating example of critical thinking gone bad, highly educated, deeply caring parents avoid the vaccinations that would protect their children from killer diseases."[46]

Before the advent of routine chicken pox inoculations for school-age children in 1995, there were about four million cases of chicken pox a year, with 100–150 deaths. Thanks to vaccination, the numbers have dropped by at least a factor of 10, and outbreaks are rare. When they occur they are often associated with anti-vaxxers.

In 2015, North Carolina state legislators tried to overturn a clause that allowed religious exemption to vaccinations. Their attempts were met with vociferous protests, and they were accused of "medical terrorism." The bill was withdrawn, and in November 2018 a chicken pox outbreak occurred at Waldorf School in Asheville. It was the largest chicken pox outbreak in North Carolina in 20 years. Two-thirds of the students in the school were not vaccinated, one of the highest exemption rates in the state.

The CDC keeps track of all cases of measles in the United States. There were outbreaks in 2014, 2015, 2017, 2018, and 2019. All occurred in under-vaccinated populations. The 2014 outbreak occurred among unvaccinated Amish communities in Ohio. The infamous Disneyland outbreak occurred in 2015, and it is estimated that about half the people who contracted the measles in that outbreak had not been vaccinated. There was also a small outbreak in 2017 a largely unvaccinated Somali-American community in Minnesota.[47] In May 2019, the MV *Freewinds*, a Scientology cruise ship, was quarantined in the Caribbean due to a measles outbreak that occurred on board, and a measles outbreak among the Orthodox Jewish community in Brooklyn and Rockland County forced officials to order mandatory immunizations for unvaccinated citizens. Larger outbreaks of measles, chicken pox, mumps, and other diseases are more likely as fewer people get vaccinated.[48]

Jennifer Reich, professor of sociology at the University of Colorado, Denver, studies the reasons parents reject vaccines for their children. They, particularly the mothers, work very hard to keep current with vaccine literature and want to do what is best for their children. "Many 'anti-vax' parents see themselves as experts on their own children, as best able to decide what their children need and whether their child needs a particular vaccine, and better qualified than health experts or public health agencies to decide what is best for their family. These decisions are inarguably not in the best interests of the community and indisputably increase risk to others who may be the most vulnerable to the worst outcomes of infection," she writes.[49] There is a lot of misinformation about flu vaccines, too. Forty-three percent of Americans have the mistaken belief that one can get the flu from being vaccinated; this is not true, as the modern vaccine doesn't contain the live virus. Reich believes anti-vaxxers are no different than most other parents who don't get the flu vaccine, unnecessarily use antibiotics, or don't turn off their cars while waiting for their children at school pickups.[50] They are just like us.

A number of studies have sought to discover how best to counter the misinformation associated with vaccines. Although these studies were done on busting flu vaccination myths, they are probably valid for all the science myths mentioned in this chapter and chapter 10. As we have seen with climate change skeptics, giving just the facts doesn't help; in fact, a backfire effect often occurs. A related approach is to try to correct an individual misconception, but in order for this to work, the correction has to be at least as interesting as the misinformation. Research has shown that if the truth is not as memorable, people who understood the correction and have no additional reasons, such as group identity, to oppose the correction, will nevertheless forget it within a few weeks and just remember the more outrageous misconception. Unfortunately, "the best evidence suggests that a more effective way of dealing with misinformation is not spreading it in the first place."[51] In large parts of the population, autism and vaccines will always be linked, despite any proof to the contrary, unless we find the cause and cure for autism.

GENETICALLY MODIFIED FOODS

Syndicated columnist Michael Gerson was appalled to find that his dog's food was prominently labeled as containing no genetically modified organisms (GMOs). "Some food companies seem to be saying that GMO ingredients are not even fit for your dog," he wrote. "These brands are guilty of crimes against rationality."[52]

A 2016 National Academy of Sciences analysis of roughly 1,000 studies led to the conclusion "that no differences have been found that implicate a higher risk to human health safety from these genetically engineered, or GE, foods than from their non-GE counterparts." The American Medical Association, the American Association for the Advancement of Science, the WHO, the French Academy of Science, and the Royal Society have all endorsed this view. In contrast, in 2018, 49 percent of the American public believed that foods containing genetically modified ingredients are worse for you than the equivalent foods containing no genetically modified ingredients; this number was an increase of 10 percent since 2016.[53] About 90 percent of scientists believe GMOs are safe. Women (56 percent) are more leery of bioengineered foods than men (43 percent). However, most Americans, in contrast to Europeans, don't think about GMOs much; their GMO beliefs are "soft" beliefs. The increase between 2016 and 2018 in public distrust of bioengineered foods has been attributed to the media creating a false balance by giving equal airtime to scientists and GMO opponents.

Horticulturists and farmers have been genetically engineering plants and animals for centuries. They interbred different species to get new and improved crops and farm animals. This was admittedly a fairly haphazard process, and the outcome was not very predictable, but genes were exchanged. A lesser known fact is that in the last few decades, farmers have been using radiation and chemical methods (both techniques are safe) to induce random mutations to supplement their crossbreeding attempts, to achieve desirable properties such as drought tolerance, color, size, and taste. As illustrated in the chapter on CRISPR (9), genetic engineering is a little different, because we use biomolecular techniques to add, remove, or change genes in an organism. Some of these changes could have occurred by crossbreeding, but others involve taking a gene from a completely different species and adding it to the genome of a new species. Two organisms with the same genome, one created by traditional crossbreeding and the other by genetic engineering, are still identical. The difficulty lies in the fact that it is impossible to prove that either is harmless. Many people are afraid that GMOs are allergenic, toxic, or carcinogenic or change the nutritional value of the food. There is no evidence of GMOs having any of these properties. Billions of animals are fed GMO feed each year, and none have shown any ill effects from the genetically engineered food.

Besides the unfounded fear of induced toxicity or carcinogenicity, there are a number of other reasons people distrust GMOs, some less legitimate than others. Perhaps the most common is a preference for nature and small family farms and an aversion to technology, especially products developed by big agricultural companies such as Monsanto. For these people the distrust of GMOs is an identity issue, similar to the climate change deniers, and they have difficulty accepting the science because it clashes with their beliefs.

Many others worry about long-term effects that haven't appeared in scientific studies yet, or that GMOs will have ecological effects that have not been observed yet. Many misguided NGOs play up these fears by using terms such as "Frankenfoods." Many producers reinforce these fears by selling "safe" GMO-free foods.

Finally, many people have the incorrect belief that natural (organic) foods are safer than foods with synthetic additives. This is wrong on two counts. First, "synthetic" and "bioengineered" are not the same, and second, natural doesn't necessarily mean less toxic.

There is a wide range of possible genetic modifications. Some are trivial, some are stupid, and some can save millions of lives. I think it is very important that consumers learn about genetic modifications and weigh the pros and cons of specific products. I would have no problem eating genetically modified foods; however, I do think I have the right to know that they have been modified.

Here is an example of genetic modifications that could save millions of lives. Evolution isn't perfect, and most crops have an imperfect photosynthetic system. In 2019, Donald Ort, the Robert Emerson Professor of Plant Science and Crop Sciences at Illinois's Carl R. Woese Institute for Genomic Biology, published a paper in *Science* in which he and his collaborators described a genetically modified tobacco plant that has a much more efficient photosynthetic system. The GMO tobacco plants grew faster and taller and produced 40 percent more biomass.[54] Tobacco was used in the studies because it is an ideal model system for other crops, such as soybeans, cowpeas, rice, potatoes, tomatoes, and eggplants. If the modification works in these crops, it is a game changer; according to Ort, "We could feed up to 200 million additional people with the calories lost to photorespiration in the Midwestern U.S. each year."[55]

TRUSTING SCIENCE

The general public distrusts scientific evidence when it comes to evolution, the safety of genetically modified foods, whether vaccines cause autism, and whether human activities are responsible for climate change. They have bought into a false picture of science. They believe that there is only 100 percent consensus or no consensus at all, that like mathematics, science has proofs.[56] As we make new findings, we publish them in peer-reviewed journals. The more pieces we have, the more of the puzzle we can see. If 90 percent of the puzzle is complete, we have a good idea of what it is showing us. There is still a huge danger in overinterpreting individual puzzle pieces and in trying to place false pieces (fake news and predatory journals) into the puzzle. But just because we don't have the whole puzzle done yet doesn't mean we don't know what it is going on. If we can get laypeople and journalists to understand that scientific research is a lot like completing puzzles, we may be able to reestablish trust in the process and results of science.

Unfortunately, the damage done by the deliberate misinterpretation of the practice of science and scientific theories is magnified in an era in which fake news is omnipresent and it is easier to not believe real news. This problem is exacerbated by the fact that most people get their news exclusively from sources whose bias they agree with, resulting in a false consensus that causes skeptics to believe that their view is more common than it really is.

Hans Rosling would like to ask all skeptics, "What kind of evidence would convince you to change your mind?" If the answer is "no evidence could ever change my mind," that shows that the deniers have placed themselves outside of evidence-based rationality and critical thinking. In that case, Rosling said, to be consistent in their skepticism about science, "they might as well ask their surgeon not to bother washing her hands next time they have an operation please."[57]

Because misleading and biased information is largely responsible for the growing distrust of science, Shanto Iyengar of Stanford University and Douglas S. Massey of Princeton University suggest, "In addition to attending to the clarity of their communications, scientists must also develop online strategies to counteract campaigns of misinformation and disinformation that will inevitably follow the release of findings threatening to partisans on either end of the political spectrum."[58] However, scientists have to be careful of being unduly harsh and dismissive. Experiments have shown that aggressive responses and comments on controversial issues do nothing but increase in-group identification.

President Trump and some of his cabinet members have aligned themselves on the side of the science deniers in all four areas I have covered here. Unfortunately, he is playing to his base: in 1974, 56 percent of conservatives expressed a great deal of confidence in the scientific community, but by 2016 that figure had dropped to 36 percent. Among Democrats there was no change.[59]

In one of his opinion pieces, Michael Gerson writes, "Our deepest beliefs should help navigate reality, not determine it."[60] He is right; in the long run, disregarding science and scientific experts will serve no political cause well.

Chapter Twelve

Quackery

Health fraud scams refer to products that claim to prevent, treat, or cure diseases or other health conditions, but are not proven safe and effective for those uses. Health fraud scams waste money and can lead to delays in getting proper diagnosis and treatment. They can also cause serious or even fatal injuries.—U.S. Food and Drug Administration

Quacks, humbugs, charlatans, and swindlers take advantage of our inability to differentiate between fact and fiction. They prey on our fear of sickness and death. It should be evident why this form of embezzlement as old as medicine itself is alive and well in today's science. Quacks are selling untested, unsafe, and very expensive fake medicines and treatments, for example, stem cell injections and health supplements. Sometimes they truly believe that what they are selling works. In the past quacks and charlatans occasionally pushed the envelope of conventionality, challenged the status quo, and were innovators responsible for some of today's medical achievements. [1] That said, I would have to be desperate before I entrusted my health to quacks of the type described in this chapter. "Bogus medical information has been circulating in one form or another since at least the Middle Ages," says Jonathan L. Stolz, a medical historian in Williamsburg, Virginia. [2]

My "favorite" quack is Robert Talbor. Although he lived more than 350 years ago (1642–1681), he was a prototypical quack, and his techniques are still used today. One of the biggest breakthroughs in the fight against malaria occurred in 1640, when it was reported that a tincture of cinchona bark was commonly used to treat malaria in South America. By 1656, the British were drinking an infusion of cinchona bark. However, they were less likely to use the treatment than the Spanish or Italians. In England, cinchona bark was known as the "Jesuit's bark," and there was some resistance to drinking the hot and bitter remedy due to its strong association with Catholicism. Talbor manipulated the situation to make a fortune. He was a self-educated doctor who saw the effectiveness of the unpopular cinchona bark brew in treating malaria fevers. He concocted a cinchona, wine, and opium mixture that was still effective but lacked the bitter taste of the original medicine and had some other interesting side effects. He sold this secret fever remedy as a safe alternative to cinchona; it was a cure untainted by Catholicism. Word of his success at treating malarial fevers spread rapidly. In 1672, Charles II appointed him royal physician, and Talbor was knighted in 1678. He became the physician of choice to royalty all over Europe. Louis XIV of France paid Talbor 3,000 gold crowns and gave him a large pension and a title, and in return Talbor promised to reveal his secret formula upon his death. The physician to royalty must have been a greedy man. He was not satisfied with his fame and fortune and clandestinely bought up all the cinchona bark to prevent any competition. Talbor's manipulations of the early pharmaceutical market did not last long. He died in 1681 at the age of 39. As promised, his secret was revealed to King Louis XIV, and in 1682 an English translation of the recipe, titled "The English Remedy: Or Talbor's Wonderful Secret for the Curing of Agues and Fevers—Sold by the Author, Sir Robert Talbor, to the Most Christian King and Since His Death Ordered by His Majesty to Be Published in French, for the Benefit of His Subjects," was published. Talbor's secret remedy was secret no more.[3] Talbor had used an existing medicine and put his own spin on it. He was a good salesman and a shrewd businessman who made and used his connections with celebrities. (I think it is okay to call a king a celebrity.) These are characteristics of quacks throughout the ages.

In the late 19th and early 20th centuries, newspapers were crammed with quacks hawking their wares, such as Mrs. Winslow's Soothing Syrup, which was sold as a cure-all for fussy babies and contained high concentrations of alcohol and some morphine. Modern charlatans no longer add opium and morphine to their products, but I am sure they would do so if they could.

By the 1920s, the American Medical Association (AMA) had managed to prevent purveyors of unproven medicines from publishing in medical journals and from advertising in newspapers. Unfortunately, the good work the AMA had done was negated by the advent of radio. By the 1930s, 45 percent of Americans had a radio, and that number rose to 90 percent by the 1940s. The quacks took advantage of the new uncontrolled media space; they hit the airwaves in a big way (sounds a bit like the advent of the internet, doesn't it?). In 1932, the Federal Radio Commission (FRC) tried to banish fortune-tellers, mystics, seers, and other people peddling dubious claims from the airwaves. However, it was not successful, and the AMA lamented that "no adequate and prompt measures are as yet available to curb venal radio stations from selling 'time' to anyone who pays the price."[4]

The first radio show host and quack to have his radio license revoked by the FRC was John R. Brinkley in 1930. The eloquent Brinkley used his radio show to get men (it's always men, isn't it?) to pay good money for goat gonads to be grafted into their testicles. This was supposed to make them as virile as goats! The AMA investigated Brinkley, finding that although he wasn't a certified doctor, he really did what he promised. The only hitch was that having goat gonads in their testicles didn't actually improve men's sex lives or increase their virility.

Shirley W. Wynne, then the New York City commissioner of health, had some advice for radio listeners. Real doctors would never promise results, because health is impossible to guarantee. Beware of things that sound too good to be true. Think critically, and be skeptical of testimonials and indiscriminate cure-alls.[5] Wynne's advice has stood the test of time and is still valid in today's internet world.

There are so many quacks in the modern world of science that they would warrant a book to themselves. In this chapter I focus on those who take advantage of the complexity of science and the good science that has been done using comparison groups, placebos, and so forth and distributed by peer-reviewed publications to at best take shortcuts in the research, and at worst make money from something untested and possibly dangerous.

SUPPLEMENTS THAT LOOK LIKE MEDICINES

Numerous supplements are dressed up to look like medicines. I focus here on Prevagen, because it is typical of the genre and because it drives me nuts that I continuously see ads for it on TV.

Most of my research is focused on a jellyfish protein, green fluorescent protein, which I introduced in chapter 5. Prevagen (sold by Quincy Bioscience) is "a dietary supplement that has been clinically shown to help with mild memory problems associated with aging." The supposed active ingredient, apoaequorin, is derived from the jellyfish I study. The product had U.S. sales of $165 million between 2007 and mid-2015, and its ads are a common sight on Fox, CNN, and NBC. You often see the ads during the evening news, on *Jeopardy*, and during sports events such as NFL games. If you watch some TV you are sure to have seen the ad (available at https://www.ispot.tv/ad/dReS/prevagen-memory-and-brain-support). It starts with a blue outline of an elderly man's head with a glowing brain enclosed in his skull. Soothing piano music accompanies a voice that says, "Your brain is an amazing thing. But as you get older, it begins to change, causing a lack of sharpness or even trouble with recall." The head turns, and we zoom in to see sparkling neurons. The voiceover continues, "Thankfully, the breakthrough in Prevagen helps your brain and actually improves memory. The secret is an ingredient originally discovered in jellyfish. In clinical trials, Prevagen has been shown to improve short-term memory. Prevagen, the name to remember." On screen the neuronal pyrotechnics are replaced by a histogram superimposed on swimming jellyfish (crystal jellies are the source of the apoaequorin). On the *y*-axis of the histogram we have 0%, 5%, 10%, 15%, 20%, 25% (no indication what this is a percentage of, but improved memory retention is implied) and on the *x*-axis we have 5, 30, and 90 days. Underneath the histogram it says, "In a computer assessed, double-blinded, placebo controlled study, Prevagen improved recall tasks in subjects" and "These statements have not been evaluated by the Food and Drug Administration (FDA). This product is not intended to treat, cure, or prevent any disease." The ad ends with an image of the Prevagen box, the pill bottle, and the logos of Walgreens, CVS, and Rite Aid, leaving the viewer with the impression that the product is endorsed by these pharmacies. I have checked my local pharmacies and Amazon; they all stock Prevagen and sell bottles of thirty pills for about $80. At this cost, the pills, which are of course not covered by any health insurance, appeal only to the desperate. Senior attorney for the AARP Foundation, Julie Nepvue, says, "Supplement companies have keyed into the idea that older people are going to spend a lot of money on their products because they want so badly to feel better and Quincy has made tons of millions of dollars on Prevagen at $33 to $60 per month."[6]

The clinical trial referred to in the advertisement was an unpublished (major red flag), dubious study of 218 human subjects doing nine different computer-assessed cognitive tasks. In January 2018, the Federal Trade Commission (FTC) and the New York Office of the Attorney General (NYAG) brought suit against Quincy Biosciences, the maker of Prevagen, for false and deceptive advertising. The complaint alleged violations of sections 5 and 12 of the FTC Act, which "prohibits unfair or deceptive acts or practices in or affecting commerce" and "prohibits false advertisements for food, drugs, devices, services, or cosmetics in or affecting commerce."[7]

Prevagen is a dietary supplement and not a drug, which is why it is not regulated by the FDA and why the FTC is trying to do something about it. However, the FTC has been repeatedly rebuffed in its efforts to regulate supplements like drugs. Perhaps we should add "if a supplement tries to look like a medicine, beware" to Shirley W. Wynne's list of ways to detect pseudoscience. In an era when the importance of facts and truths has been devalued, it is not surprising that the makers of Prevagen and its ilk are able to continue advertising and selling fake goods.

Since 2008 the FTC has filed 120 cases challenging the medical claims for supplements. Products claiming to slow down or even undo the mental ravages of aging and weight-loss supplements are the most common offenders. Joseph Jankovic, a neurologist who heads the Parkinson's Disease Center and Movement Disorders Clinic at Baylor College of Medicine in Houston, says fake supplements for Parkinson's are a "major, major problem, I have to spend a lot of time in patient visits debunking unreliable information from the internet."[8]

STEM CELLS

Stem cell clinics are popping up all over the United States. They have grown from just 2 in 2009 to over 700 in 2018. Since 2014, at least 100 new facilities have opened in the United States every year.[9] They are following on the heels of many stem cell clinics that were established in countries like China, South Korea, and Mexico, where they target Americans as "medical tourists." A 2019 study of medical marketing in the United States shows that stem cell clinics boosted their marketing from $900,000 in 2012 to $11.3 million in 2016.[10]

The two most important properties of stem cells are their abilities to undergo self-renewal and to differentiate into new cell types. They can multiply to form many daughter cells, which can then change into (differentiate) new cell types as required. Stem cells derived from embryos are called "embryonic stem cells." They are tremendously good at replicating and can be differentiated into any type of adult cell. Stem cells obtained from adults are less good at dividing and can only differentiate into a limited number of cell types.

In regenerative medicine, stem cells are guided to replace damaged or diseased tissue. Stem cell research has now reached the point where the first carefully regulated human trials are being done. Stem cell therapies in spinal cord injuries, type 1 diabetes, Parkinson's disease, amyotrophic lateral sclerosis, Alzheimer's disease, heart disease, stroke, burns, cancer, and osteoarthritis are all being considered. The problem is that unscrupulous stem cell clinics are undermining the field by skipping ahead without doing efficacy or safety studies, and they are selling spots in their so-called medical trials.

The FDA has issued the following warning: "Stem cell products have the potential to treat many medical conditions and diseases. But for almost all of these products, it is not yet known whether the product has any benefit—or if the product is safe to use."[11] The only FDA-approved stem cell–based products are blood-forming stem cells derived from umbilical cords. Stem cell therapies have numerous dangers: the stem cells could differentiate into the wrong cell types, or into the right cell types but nonfunctioning; they can grow irregularly, induce an immune response, or cause a tumor growth. Consumers should only consider FDA-approved treatments or studies being done under an Investigational New Drug Application (IND), which is a clinical investigation plan submitted to and allowed to proceed by the FDA.[12]

My father is 91 and has age-related macular degeneration (AMD). He can still read, but that will become harder and harder. Macular degeneration mainly affects people over age 50. It is the most common ocular disease in the developed world, causing more than 50 percent of all cases of visual impairment. The disease affects cells in the macula, which is located in the center of the retina in the back of the eye. Researchers from the University of California, Santa Barbara, derived retinal pigment epithelium cells from some stem cells and collaborated with surgeons from Moorfields Eye Hospital, London, who used a specially engineered surgical tool to insert the patches of epithelial cells under the retinas of two patients in a very small clinical trial. The operation lasted one to two hours and was a great success. An 86-year-old man and a woman in her early 60s went from not being able to read at all, even with glasses, to being able to read 60 to 80 words per minute with normal reading glasses. [13] "This study represents real progress in regenerative medicine and opens the door on new treatment options for people with age-related macular degeneration," says University of California, Santa Barbara, coauthor Peter Coffey. [14]

Doing research of this nature is a slow, methodical process, regulated by the FDA and its international counterparts. Small numbers of patients are treated at first, only one eye is done, there is a comparison group, and there is a long follow-up. In this case, the paper was written after the two subjects had been observed for 12 months after the operation. It will be years before this treatment has been rigorously tested and can be safely taken to market.

Stem cell clinics are unregulated and use false results or legitimate early results such as those described in the preceding paragraph to attract patients for stem cell replacement surgeries and circumvent the FDA approval process. They use high-pressure sales techniques to recruit their patients, including on-the-spot discounts, emotional testimonial videos, and slick recruitment videos. It is more like buying a time-share condominium than a doctor's appointment. Surgeries range from $2,000 to $20,000. Patients who can't afford the fees are advised to start GoFundMe pages and to get the money from family and friends. [15] A study published in June 2019 revealed that more than half the stem cell clinics studied employed no physicians trained to deal with the conditions advertised. [16]

The Stem Cell Center of Georgia is located in the Ageless Wellness Center in suburban Atlanta. It was founded in 2008, specializing in Botox treatment, laser hair removal, and other beauty treatments. In 2014, the center added stem cell treatments to its menu. Doris Tyler was losing her sight, and in 2016 she signed up for stem cell treatment of her macular degeneration at the center. The clinic staff claimed they could take stem cells from her fat and inject them into her eyes, where they would halt or even cure the macular degeneration. Hers was the first case in which the clinic tried this procedure, and five days after the operation staff boasted of its success on their Facebook page and urged customers to book appointments. But Tyler's vision was becoming blurry and her retina detached, and despite numerous corrective surgeries she totally lost her vision in a few months. Doris Tyler's husband is suing the clinic.[17] Three other women have lost their sight in similar procedures done at a Florida clinic, the US Stem Cell Center. This center has since settled the three cases and has stopped doing eye injections. Nevertheless, it has expanded by opening a new facility in a central Florida senior living community and has added stem cell treatments of erectile dysfunction to its procedures (remind you of the goat gonad grafts done by John Brinkley in the 1920s?).

Stem cell clinics are not regulated by the FDA, as they obtain their stem cells from the patients themselves, and therefore the stem cells can't be considered new drugs. However, the FDA has promised to crack down on stem cell clinics, particularly those that are performing dangerous operations such as injections into the eyes, spinal cord, or brain. The clinics and their doctors cite testimonials of "cured" patients as evidence that their treatments work and say that customers should be allowed to sign up for experimental treatments, claiming it is their right. Timothy Caulfield, a health law professor at the University of Alberta, disagrees: "What they're really selling is false hope. It's scienceploitation. They're taking a legitimate and developing field of science and using it to prey on patients who are desperate for a cure."[18] Charles Murry, director of the Institute for Stem Cell and Regenerative Medicine at the University of Washington, is "afraid that these charlatans will besmirch the reputation of legitimate work we have spent decades trying to bring to the clinic."[19]

Scientists are always finding new, promising leads that could open the door to a revolution in medicine. Mostly they end up being dead ends that don't translate from mouse studies to human treatments or have undesirable side effects that make them unusable. That doesn't bother quacks; they have found ways of hitching their unscrupulous get-rich schemes to legitimate research breakthroughs as they follow the path from promise to discarded dream.

PARABIOSIS

In 1864, Paul Bart removed the skin from the sides of two albino rats and stitched the two rats together so they shared a circulatory system, as found in conjoined twins and between a mother and her unborn child. He injected fluid into one rat and showed that it flowed into the other. This was the first reported joining of the vascular system of two separate living organisms, and so parabiosis was born (in Greek "para" means between, and "bio" means life). The field underwent a significant resurgence when scientists showed that young blood rejuvenates old tissues (in mice). Clive McCay, a professor of animal husbandry at Cornell University, was the first to study the effects of parabiosis on the aging of rats. In 1956, his students joined 69 pairs of young and old rats. It wasn't a great success; 11 pairs died from a mysterious condition the researchers called "parabiotic toxicity," and they found that "if two rats are not adjusted to each other, one will chew the head of the other until it is destroyed."[20] Later studies by the same group showed that a shared circulation increased the bone density of older rats. In the 1970s, other groups showed that in some cases shared vascular systems could lead to prolonged life for the older partner, but the field soon fell out of favor as institutional approval for animal studies became more demanding. That has all changed in the 21st century, as potentially promising results have made it worthwhile to socialize the mice prior to parabiosis and join only mice of the same sex and size. By connecting the circulatory systems of young mice with those of old mice, researchers have made the old mice healthier, smarter, and stronger; even their fur is shinier. Initial studies from Amy Wagers's laboratory at the Harvard Department of Stem Cell and Regenerative Biology have shown that the growth differentiation factor 11 in the blood of young mice may be responsible for the increase in strength and stamina observed in old mice after parabiosis. "We're not de-ageing animals," says Wagers. "We're restoring function to tissues."[21]

In 2014 Tony Wyss-Coray, a neurologist at Stanford University, and his collaborators published a paper in *Nature Medicine* in which they reported that young blood plasma stimulated neuronal growth in old mice.[22] "We didn't have to exchange the whole blood," says Wyss-Coray. "It acts like a drug."[23] Based on their results, the researchers formed a biotech company, Alkahest, which is conducting randomized, placebo-controlled, double-blind clinical trials on the effect of parabiosis on Alzheimer's and Parkinson's disease. The company is not interested in starting up a clinic; the aim is to develop drugs for age-related diseases. Tony Wyss-Coray says they're focusing on drugs because "there's just no clinical evidence [that the treatment will be beneficial], and you're basically abusing people's trust and the public excitement around this."[24]

Of course that hasn't stopped quacks and charlatans. In 2019, *Men's Health*, which is not a peer-reviewed journal, published "People Are Getting Transfusions with Young People's Blood to Fight Aging."[25] The article is about Ambrosia, a company that for approximately $12,000 (it was $8,000 in 2018) will infuse a person with one to two liters of plasma taken from some 16- to 25-year-old donors. The procedure takes one to two days. Dr. Jesse Karmazin, founder of the company, says they have done about 150 transfusions, and many of his patients report feeling better and have improved memory, sleep, and appearance with just one treatment a year. Dr. Karmazin claims he has done a preliminary study of 81 patients, who (you guessed it) paid for their own treatments. Supposedly the study showed improvement in markers for Alzheimer's, heart disease, and inflammation, but it has not been published. According to Ambrosia's website, you can schedule treatments in Phoenix, Los Angeles, San Francisco, Tampa, Omaha, and Houston, and the company accepts payment via PayPal. A New York clinic is in the plans. Because these clinics are essentially doing blood transfusions, the procedures are automatically approved by the FDA, and Ambrosia doesn't have to prove that its treatment has any benefits.[26]

Dr. Dipnarine Maharaj, a Scottish-trained hematologist and oncologist, has upped the ante. He is selling spots in his clinical trials for $285,000. What do you get for your money? Monthly infusions of plasma obtained from young donors who have been injected with a drug designed to stimulate the bone marrow to make more white blood cells and stem cells. Rebecca Robbins, life science reporter for *STAT*, asked eight independent experts to review Maharaj's informational handouts about the clinical trial, and all sharply criticized the studies' scientific rationale.[27] Michael Conboy, a cell and molecular biologist at the University of California, Berkeley, who has done parabiosis studies, said, "It just reeks of snake oil. There's no evidence in my mind that it's going to work."[28] Robbins asked Maharaj for the scientific basis of his expensive clinical trials, and Maharaj provided her with six papers he asserted were fundamental to his work. Amy Wagers, who is a coauthor on three of the papers, stated in an email to *STAT* that she does not agree that her teams' studies provide any scientific basis for Maharaj's clinical trial.[29]

DR. OZ

Kelly McBride, vice president of the Poynter Institute, a nonprofit journalism school in St. Petersburg, Florida, thinks that fake medical news is worse than any other type of fake news, and there is more of it out there. A lot of it is due to quacks and charlatans who want to make a quick buck, as illustrated by the stem cell and parabiosis examples here. But there is also a substantial amount of fake news generated by journalists and TV hosts, who according to McBride struggle with the facts that "journalism is very much about trying to simplify and distribute information about what's new and where advances have been made. That's incompatible with the scientific process, which can take a long time to build a body of evidence" and that "good information can be boring."[30]

The Dr. Oz Show has an estimated weekly audience of 35 million viewers in 215 markets around the world. Dr. Mehmet Oz got his start on *The Oprah Winfrey Show* and was so popular that he got his own show in 2009. Both *Time* and *Esquire* have included him on their lists of the 100 most influential people. It is difficult to keep up that profile and enthrall that many viewers for 10 years. A 2014 study published in the *British Medical Journal* (*BMJ*) examined the recommendations made in 40 randomly chosen episodes of *The Dr. Oz Show* from 2013. The authors found that 13 percent of his recommendations were completely contradicted by the prevailing peer-reviewed medical literature, and that there was no evidence for 39 percent of his assertions.[31] The *BMJ* paper is not the first peer-reviewed article to call out Dr. Oz by name. In 2013, the journal *Nutrition and Cancer* published a paper titled "Reality Check: There Is No Such Thing as a Miracle Food."[32] It was written to warn doctors and patients that statements Dr. Oz was making about "miracle foods" that could decrease the risk of ovarian cancer had no medical evidence to back them up. Oz has also been castigated by a Senate subcommittee on consumer protection and the FTC for using his show to deceive audiences and sell products, particularly diet supplements: belly blasters and mega metabolism boosters. He does not disclose any conflicts of interest or commercial interests he has in products he endorses on his show. In response to criticism, Dr. Oz has invoked his right to free speech and claimed there is a conspiracy by the pro-GMO lobby to undermine him; he also said, "I want folks to realize that I'm a doctor, and I'm coming into their lives to be supportive of them. But it's not a medical show."[33] A show called *Dr. Oz*, in which he wears medical scrubs and dispenses medical advice, is not a medical show? After all the complaints, petitions, and subpoenas, Dr. Oz still has anti-vaxxers on his show; in 2018 he aired an episode titled "What Your Astrological Sign Can Tell You about Your Health"; he pushes miracle elixirs, homeopathy, and imaginary energies; and he touts psychics who communicate with the dead. He regularly uses words like "miracle" and "magic." And yet he is still "America's Doctor"; his audience keeps growing, and they believe him. Despite all the bad publicity and his obvious quackery, Oz won a daytime Emmy in 2018, and his show aired in China in 2019. In May 2018, President Trump appointed Dr. Oz to his council on sports, fitness, and nutrition. He has written six books and has his own magazine.

WELLNESS

Dr. Oz is a big proponent of "wellness." In 2017, the wellness industry made more than $4.2 trillion; the personal care, beauty, and antiaging segment of the industry made $1.083 trillion by itself, more than the pharmaceutical industry.[34] Wellness is a difficult concept to define; the WHO's definition, which is often used, states: "Wellness is a state of complete physical, mental, and social well-being, and not merely the absence of disease or infirmity." It is about feeling good, being happy and healthy, doing yoga, going to spas, and walking on the beach. Sounds pretty good, doesn't it? And I think most of it is pretty good, but there is just too much money and fame involved for it all to be good. Wellness isn't medicine, but it fills some gaps left by medicine, and its popularity may be a reaction to our hectic, stressful modern lives and ironically, against the commercialization of medicine.

Women are the primary target of the wellness marketeers. Doctors often don't have or take the time to listen to patients, particularly women, when they talk about their health issues. For example, researchers from the University of Florida and the Mayo Institute monitored conversations between doctors and patients and found that patients were given an average of 12 seconds to explain why they were seeking treatment before they were interrupted.[35] "Even if the clinical advice is sound the idea of not being listened to matters. And there is evidence that women's issues aren't taken as seriously. Their problems aren't being heard. That matters. Those are real problems," says Timothy Caulfield.[36] In order to understand why people are attracted to wellness scams and to get material for his book, titled *Is Gwyneth Paltrow Wrong about Everything? When Celebrity Culture and Science Clash*, Caulfield visited many wellness practitioners. He reports, "I've gone to a lot of different alternative practitioners: reflexologists, acupuncturists, reiki, cupping. Almost without exception, it's been a positive experience. Someone is listening to your problems. With traditional healthcare, it's always a pretty miserable experience."[37] Both genuine wellness practitioners and wellness quacks are there to fill the void created by a stressed medical care system. They also take advantage of the fact that science, especially medical science, has historically neglected women. When it comes to women's health, "There's a lot we don't know," says Amy Miller, president and CEO of the Society for Women's Health Research. "We only started including women in research 15 or 20 years ago and that means a lot of generic drugs may not have been investigated in women. We don't know if a drug isn't as effective in a woman's body as it is in a man's. And then there are areas where there is nothing on the market [for women]."[38] It wasn't until 2016 that the National Institutes of Health (NIH) required all the researchers it funded to use both male and female mice in their studies and to use male and female tissue cells in their research. Perhaps it is understandable then that people, especially women, are looking elsewhere for their medical solutions and are vulnerable to pseudoscience.

The title of a 2019 article in *Marie Claire*, "The Wellness Industry Isn't Making You Well: It's Exclusionary, Expensive, and Too Trendy for Its Own Good," sums up some of the problems with the industry. However, I am more interested in how science is being perverted to make a profit and how both science and the patient may be damaged in the process.[39] Wellness appears in this chapter because "in its current form, wellness isn't filling in the gaps left by medicine. It's exploiting them."[40]

Activated charcoal is a good example of how the wellness industry is using legitimate science to make lots and lots of money in a manner similar to that seen in stem cell and parabiosis clinics. Chemists use activated charcoal as a filter to remove contaminants and impurities from solutions. The technique relies on the fact that activated carbon has an extensive surface area, which can adsorb pollutants. Emergency room doctors use activated carbon to treat patients who have been poisoned. As the activated carbon passes through the gut, it binds all the nonpolar materials in the gastrointestinal tract, good and bad, which are expelled in dark-colored stools. That is a good thing if you have just swallowed some poisons by accident. It will not cure hangovers, detoxify you, reduce bloating, prevent aging, or improve your heart health, no matter what you see on *The Dr. Oz Show* or read on health blogs. I have to admit that even I am a little tempted to try activated charcoal when I read statements like this one, taken from the *Bulletproof* blog: "Travel, environmental pollution, and low-quality food can all put toxins in your system—and activated charcoal is the perfect solution when that toxic load gets too heavy to bear. From binding poisons to eliminating body odor, activated charcoal is loaded with benefits that can help keep toxin-induced fog and fatigue at bay."[41] I am a chemist, so I know better; I am well aware that activated charcoal will indiscriminately bind all nonpolar molecules in my gastrointestinal tract (including antioxidants, medicines, and vitamins), and that it won't cross over into my circulatory system or bind to toxins in my blood before miraculously coming back into my gut and forming some toxin-loaded, licorice-colored stool. Yet due to a psychological phenomenon called the illusory truth effect, my knowledge about the charcoal doesn't necessarily protect me. Even a single exposure to information that sounds quasi-plausible—and there is often enough science in these scams to sound slightly real—can increase the perception of accuracy and tempt me to purchase the activated charcoal smoothies and vegan charcoal lattes advertised on Dr. Oz's website.

CONCLUSION: BEWARE UNSCRUPULOUS QUACKS SELLING PSEUDOSCIENCE

The complexity of science, the broad reach of social media, and the abundance of misinformation have made it easy for unscrupulous quacks to sell pseudoscience. Besides bilking their victims of money, quacks foment scientific uncertainty and undermine the public trust in science and scientists. Their call for people to sign up for new, unproven techniques or to try "something natural" can lead people with serious illnesses to postpone effective medical care. Both the FDA[42] and FTC[43] websites have pages with regularly updated health fraud news. The FDA's website warns readers that "testimonials" are not a substitute for scientific proof; to be wary of promotions using words such as "scientific breakthrough," "miraculous cure," "secret ingredient," and "ancient remedy"; and to be skeptical of claims that the product is "natural" or "non-toxic" (which doesn't necessarily mean safe). Beware of celebrities selling "medicines" or diet supplements and offering medical advice, and remember that not all chemicals are bad and not everything natural is good.

Quacks often appeal to American notions of freedom and individuality, a resentment of wealthy doctors, and a desire to stay up to date with the newest trends. They make unscientific claims and require society to disprove them, while doctors use proven methods in their treatments. It is very difficult to prosecute even the most flagrant quackery. To prevent it, we need to promote scientific knowledge and critical thinking and to curtail fake news.

Part Six

Future Science

Chapter Thirteen

Some Thoughts on Keeping Science Healthy in a Sustainable Future

We are in charge of our own evolution and the health of the planet; that is a lot of responsibility. With increasing scientific ability, the question posed by Dr. Jonas Salk, the inventor of the polio vaccine—"Are we being good ancestors?"—becomes ever more relevant.

THE ROLE OF SCIENCE IN STOPPING CLIMATE CHANGE

Climate change is the biggest challenge facing our generation, and I don't see science solving this problem. I grew up in South Africa. My childhood dream was to be a game warden. That dream was shattered when I failed botany in my freshman year (it clashed with antiapartheid protest meetings at lunch). I am often amazed by friends who have gone on to become game wardens. They believe that climate change is real, and that it is caused by fossil fuel combustion; however, they are not particularly concerned. They have misplaced confidence in science. Like many others, my old university friends believe that scientists will find a painless solution. I don't think that will happen. Scientists cannot tell us what to do about climate change. Drastic social solutions are required. Science will help us develop and improve electric autonomous cars, renewable energy sources, meat alternatives, carbon dioxide sequestration, and so forth. But such inventions are not enough; the solution requires a change in the way we live. We will have to build smaller houses, travel less, eat less red meat, and throttle back our consumption.

To make the changes required to prevent catastrophic climate change, we are going to have to change the minds of both climate change believers who have too much confidence in science and climate change deniers who don't trust climate scientists: two very different audiences. That is going to require a lot of faith in science and scientists at a time when there is not much faith going around and truth is negotiable. I am not the only one concerned about this. The decline in trust "terrifies" former MIT president Susan Hockfield, who says we have no way of making it into the future if we don't trust scientists to be experts in their fields. On the podcast *Recode Decode with Kara Swisher*, Hockfield urged, "We have to insist on an understanding that there are people who understand areas better than we do. I don't pretend to be an engineer. I don't pretend to be a physicist. If the physicists at MIT tell me that they've figured out gravitational waves, I'm going to trust them more than I'm going to trust myself to imagine whether or not there are gravitational waves." She added that she understands "that people might debate the fine points of climate change, but the fact is that the best science indicates that we're in trouble."[1]

In May 2019, New Zealand became the first country to announce a national budget that placed the "well-being" of its citizens over economic growth. According to Bill McKibben, this would be a sign of a mature economy, an economy that has reached its adult size and no longer needs to grow. To stop climate change, we will all have to trust the science enough to realize that we are all grown up and that it is time to stop growing. We don't need new malls, housing developments, bigger cars, and so forth. We have to accept that our teen years are over and we have to live responsibly and, in the interest of our communities, place limits on our behaviors. In *Falter: Has the Human Game Begun to Play Itself Out?*, McKibben exhorts us to "imagine a society interested more in economic and political maintenance and contentment than in exhilaration and extension."[2] It may not be the American way, but perhaps machine learning and AI can help us move to a "well-being" economy. If the advances in these fields result in the feared reductions in jobs (e.g., of truck drivers, medical doctors, and those in service industries), perhaps we can "make lemonade out of lemons" by redistributing responsibilities, reducing our working hours, and focusing on upkeep and sustainability rather than growth—that is, move from a growth economy to a mature economy.

Interestingly, although China is much more populous and in a much steeper growing phase than the United States, I think because it is an autocratic state it will be able to make the changes required to halt climate change easier and faster than the United States. Julian Savulescu, an Australian philosopher and bioethicist at the University of Oxford, echoes these thoughts; he thinks democracies can't solve climate change, "for in order to do so a majority of their voters must support the adoption of substantial restrictions on their excessively consumerist lifestyle, and there is no indication they would be willing to make such sacrifices."[3]

No matter how optimistic/pessimistic they are, all climate scientists agree that climate change is real and is caused by humans. Something drastic has to be done, and the only way to convince politicians and the electorate to make the required changes is to restore the public trust in science and the scientific process.

TRUSTING EXPERTS—AND THE TRUMP ADMINISTRATION

We all need experts. Scientists will be happy to tell you that journalists, consumers, and policy makers don't understand science and need expert advice. But they themselves struggle to see that they also need input from experts. We scientists need to learn how to increase the diversity of our field, consider the ethics of what we are doing, and reach out to nonscientists and explain our work.

The knowledge gap between experts (be they scientists, medical doctors, or economists) and nonexperts is increasing, and perversely our reliance on these experts is decreasing. This is due in part to the fact that we often have a misguided overconfidence in our knowledge, something Leonid Rozenblit and Frank Keil, psychologists at Yale University, termed the illusion of explanatory depth. According to them, "most people feel they understand the world with far greater detail, coherence, and depth than they really do"[4] and consequently have less use for experts than they should.

A little more than a century ago, Richard Burdon Haldane wrote a report for the British Parliament in which he, a career politician, advocated putting scientists in charge of deciding who should receive government research funding and that politicians should stay out of funding deliberations and when making science policy should listen to the advice of experts. His ideas would come to be known as the Haldane Principle, and it would take the upheaval of the Second World War for them to take root.

In World War II, scientific technologies such as radar and the atomic bomb were central to the Allies' victory. Following the war, scientists advocated for a separation between science policy and politics, arguing that if science could win wars, it could also help to hold peace. They were successful, and many nations adopted the Haldane Principle. By 2015, U.S. federal agencies had about 1,000 advisory panels relying on the expertise of roughly 60,000 members. They advised government agencies on everything from foodborne illness, to drug safety, to air and water pollution. They were there to provide fact-based advice and evidence. After all, no one can reasonably expect our elected representatives to master every topic; they have to rely on the expertise of professionals.

However, recently that approach has changed. On June 14, 2019, President Trump signed an executive order to reduce all advisory boards by one-third, a telling move that will have long-term consequences.[5] In a comment published in *Nature* in 2018, Ehsan Masood worries that an expanding global network of populist political movements is deriding independent scholarship and erasing the protections granted by the Haldane Principle. Masood calls out Trump and the UK Brexiteers and worries, "Today, from Istanbul to Islamabad, from Rome to Rio de Janeiro, a parade of authoritarian leaders is advancing policies that fly in the face of evidence—on energy, emissions, the environment, economics, immigration and more. Worse, these leaders are demanding that academics march to the beat of their drums."[6] Scientists are no longer advising politicians, and politics is driving science policy.

This is particularly true for the Trump administration, which in making its science policy has been much more interested in soliciting advice from industrialists than scientists. It has scant regard for the Haldane Principle. It has at best an apathetic stance toward science and at worst is actively hostile, particularly toward the environmental sciences. In the first year of the Trump administration, nearly two-thirds of the science advisory boards met less often than required by their charters, and they met less frequently than in any other administration. From the beginning, the administration has tried to reduce the staff and budget of the Environmental Protection Agency (EPA) and to undermine its use of science in policy making. As a consequence, there have been steep drops in enforcement of the Clean Air Act and Clean Water Act in nearly all regions of the United States.[7] And environmental policies have been amended on advice from industrial advisers against the advice of the scientific community. There are so many instances of the Trump administration attempting to scale back or wholly eliminate federal climate mitigation and adaptation guidelines that I can't catalog them all. The Columbia Law School's Sabin Center for Climate Change Law maintains the Climate Deregulation Tracker (http://columbiaclimatelaw.com/resources/climate-deregulation-tracker/).

In July 2019, *Science* reported that "FDA [Food and Drug Administration] enforcement actions plummet under Trump." The FDA is one of the nation's critical watchdogs. Its warning letters, its primary tool for keeping dangerous and ineffective pharmaceuticals off the market, have dropped by a third. The agency has been particularly lax in its oversight of medical devices, with the Center for Devices and Radiological Health sending out two-thirds fewer warnings.[8] The danger in all these attacks on science is that it is nearly impossible to keep track of them. We are so bombarded with blatant misrepresentations of science and environmental deregulation that we are overwhelmed, and some pass us by.

In *The Workshop and the World: What Ten Thinkers Can Teach Us about Science and Authority*, Robert Crease tries to find lessons in the writings of ten philosophers that can be used to combat the erosion of scientific authority.[9] Many of his ideas are presented in this book. In the book he compares

U.S. politicians who attack science to ISIS militants who blow up archaeological treasures and smash statues. After all, he argues, science is a cornerstone of Western civilization.

Philip B. Duffy, president of the Woods Hole Research Center and a National Academy of Sciences panelist, who reviewed the government's most recent climate assessment, told the *New York Times*, "What we have here is a pretty blatant attempt to politicize the science—to push the science in a direction that's consistent with their politics." He added, "It reminds me of the Soviet Union."[10] That is not what we want to hear from an impartial scientist. The Trump administration is certainly not the first administration to change the scientific policies of its predecessors, as is its right. However, it is the first to systematically chip away at large portions of science-related policies and legislation without consulting scientists, while at the same time undermining the scientific process to discredit climate scientists and public health researchers. It is difficult to accept that Trump, whose errors in science are so crude and blatant, can undermine science. How can someone who says that the noise windmills make causes cancer[11] command more respect than the combined thinking of thousands of scientists? But in today's politics, anything is possible, and people are happy to accept his inane utterances over the well-proven scientific method.

To me, nothing demonstrates the intention to influence the way we think about science more than the fact that employees of the Centers for Disease Control are now not allowed to use the words "evidence-based," "science-based," and "fetus," and that staff at the United States Department of Agriculture have been told never to use the phrase "climate change"; instead they have to say "weather extremes" and replace "adaptation" with "resilience." Soviet indeed. The devaluation of science by the Trump administration has had some unfortunate consequences. John Holdren, who was President Obama's science adviser and may be slightly biased, has said, "Trump, along with his picks to lead major scientific agencies, act as talent repellents, because they appear to know little and care less about the value of insights from science and technology in shaping public policy."[12]

It is important to remember that Trump's disregard of science and the environment plays well with a substantial fraction of the U.S. population and doesn't reflect just the opinion of one person. A significant segment of the population resents science.

The most obvious future consequences of the undermining of the scientific process and the marginalization of science, particularly public health and environmental science, by the Trump administration are difficulties in dealing with climate change, the degradation of the U.S. environment, and a decline in the competitiveness of U.S. science and technology.

THE SCIENCE RACE AND U.S. COMPETITIVENESS

American research universities and institutes have been at the forefront of scientific research since the end of the Second World War. According to Susan Hockfield, when she was president of MIT hardly a week would go by "when someone from some other country wasn't in my office saying, 'We understand what the United States did. We want to do it in our country. Can you help us understand how we can build something like MIT because we understand that is part of the recipe to build an economy like the United States has enjoyed?'"[13] I think we can safely say that those days are disappearing; America's scientific preeminence is now dwindling. In the 2010s, the world's economic center of gravity, which is calculated using the average country's location weighted by its gross domestic product, has steadily shifted from the United States and Europe toward China. There is no world scientific center of gravity, but if there was one I am sure it would also be migrating toward China. AlphaGo might have been responsible for China's AI Sputnik moment, but China didn't need an ignition of its science programs; they were already in high gear. China has overtaken the United States in total number of science publications, and in December 2019 China was named the biggest producer of high-quality research in chemistry by the Nature Index, knocking the United States into second place. This is the first time China has been in this position.[14]

In 2000, 5 percent of global spending on research and development was by China, and the United States accounted for 40 percent. Just 15 years later, China was responsible for spending 21 percent of the global R&D funding, with the United States spending 29 percent.[15] Since then the Trump administration has proposed cutting science funding three years in a row. Fortunately, so far Congress has restored the funds. A study of over 10 million U.S. patents and scientific papers published between 1926 and 2017 found that U.S. patents have increasingly relied on federally funded science. Today at least a third of all U.S. patents have their roots in federally funded research.[16] Federally funded science keeps this country competitive; it would be a big mistake to cut it.

While at MIT, Hockfield saw ample evidence of China's interest in science:

> When our Chinese colleagues came with the same question, they were enormously insightful because many of the others would say, "We just want to know how to build engineering. We need engineering departments because that's where the tech transfer comes from. We understand that's how some technology gets developed." But our Chinese colleagues would say, "We get it. We're going to build the best physics department, we're going to build the best math departments, we're going to build the best biology departments.

We're going to do basic science because we understand that's at the foundation of the engineering of the technologies of the future."[17]

Hockfield is still optimistic; she thinks the United States will retain its position at the forefront of the scientific world because of our insistence on scientific integrity and competitive instincts.

I am not so sure. Our scientific integrity may be overrated, there is a lot of irreproducible science out there, and our scientific production relies heavily on foreign scientists. Can we expect scientists to truthfully report their results, warts and all, when the stakes—funding, tenure, and promotion—have increased, fake news is ubiquitous, and the media and president have a fairly relaxed definition of fact? And can we expect more foreign scientists to come here as our immigration laws get tighter and the social climate is becoming less accepting of outsiders? As chapter 2 points out, fewer international graduate students (the engine of our academic research system and the talent pool that feeds our industries) are coming to the United States. According to the Department of Homeland Security, there was a 4 percent drop in international graduate students coming to the United States between 2016 and 2017, caused in part by 18,590 fewer international graduate students enrolling in computer science and engineering.[18]

At the same time, Chinese Americans working in the United States are being placed in a very difficult position. Their situation is perhaps best expressed by Yangyang Cheng, a postdoctoral research associate at Cornell University, in an article titled "My Science Has No Nationality—Chinese Scientists in the Age of Trump." She asks: "For Chinese scientists who immigrated to the U.S., where do their hearts and bodies belong? In their home country, where an authoritarian government is increasing its hold on society, aided by technology for surveillance and censorship? Or in a country whose president actively rejects them, where they are painted as spies?"[19]

Growing political tensions between the United States and China, a trade war, and a crackdown on foreign scientists (read Chinese) have led to a great amount of unease among Chinese American scientists. The clampdown on international scientists has included reduced access to visas and a government-instigated investigation of foreign entities for interfering in the funding, research, and peer review of National Institutes of Health. This investigation has led to the dismissal of five Chinese researchers for "sharing grant proposals that they were reviewing, and for failing to disclose foreign funding and affiliations at institutions abroad." This is a bit like jailing someone for plagiarism.[20] The situation compelled MIT president L. Rafael Reif to write a letter to the entire MIT community expressing his dismay at the situation. In it he asserted that MIT and the United States have flourished because MIT has been a magnet for the world's finest talent, who act as a kind of oxygen energizing the institution.[21]

Our xenophobia is interfering with our scientific progress and limiting our scientific competitiveness. Yangyang Cheng feels that "on the proxy battleground for technological might, scientists are spoken of as strategic assets, instead of human beings with desires and agency." This can't go well.

For many Chinese, going back to China is an increasingly attractive option. This would be a massive loss for U.S. science. Just consider Feng Zhang, who has been instrumental in the development of both CRISPR and optogenetics. He is truly a science superstar. He was born in China and came to Iowa at age 11. Imagine if his parents hadn't come to the United States or if he leaves for Switzerland. What a blow that would be to U.S. science! Now multiply that by the thousands of international graduate students who are not coming here.

In May 2019, President Trump introduced a merit-based immigration system. This is not the answer. If we had had a merit-based points system of immigration in the past, people like Irving Berlin, Alexander Hamilton, Andrew Carnegie, I. M. Pei, and Ralph Baer, the inventor of the game box, would never have been allowed to enter the United States. Many of our most important scientists came to America as children with parents who would not have qualified for entry: they didn't have big bank accounts, impeccable English, or plans for a new factory in their back pockets; they were molded by America just as they helped mold it. [22]

If the United States wants to stay competitive with China and the rest of the scientific world, it has to strike a very careful balance between basic and applied research and between ethics and growth. Who is going to make these choices? The amount of scientific knowledge is rapidly increasing. This has resulted in an ever-widening gap between the scientific knowledge of legislators, religious leaders, and voters, and the total available science knowledge. Recall from chapter 1 that scientists are underrepresented in legislatures, being outrepresented by radio hosts, farmers, and even ordained ministers. This means that science legislation, funding, and direction are being determined by an increasingly divided Congress that does not have the requisite scientific knowledge to do the work and sometimes doesn't respect the experts. But it is the advice of experts legislators are going to need.

To maximize our scientific potential, we need to get our best minds working on important problems, and then we have to listen to them. Always remembering that we can never predict where the next scientific breakthroughs will come from, we need to attract and retain scientists from all nations, genders, and creeds.

SHOULD THERE BE LIMITS ON SCIENCE?
IF SO, WHO SETS THEM?

In 1926, J. B. S. Haldane, the British statesman whose uncle was Richard Burdon Haldane, wrote the essay "On Being the Right Size," in which he said that if we drop a mouse from a building it survives, "a rat is killed, a man is broken, a horse splashes." Gravity limits our size; it is the enemy of the large. Elephants have massive hearts to pump oxygen-bearing blood throughout their bodies, and they have thick, sturdy bones to remain upright. The larger the organism, the higher its complexity; just compare a flea with a giraffe. But complexity only gets us so far; to get larger we have to change the conditions, such as going from a terrestrial to a marine environment. Paul Saffo, professor of mechanical engineering at Stanford University, argues that Haldane's observations apply to more than just organisms: "Everything from airplanes to institutions has an intrinsic right size, which we ignore at our peril."[23] What about science? What is its right size?

Science and technology have enabled robust growth for humans going back to our earliest origins. But can we let that growth continue unchecked? Many scientific breakthroughs will benefit humanity, but their advantages are often offset by potential abuses. These abuses need to be limited when they are dangerous and when they exceed a commonly established ethical boundary. I am particularly concerned about developments in CRISPR, optogenetics, and artificial intelligence. How do we limit growth, especially when the limits may curtail our competitiveness?

With CRISPR, especially when it is associated with a gene drive, we will soon reach the point where for the first time in history one person or one experiment can potentially wipe out a whole species, including humans themselves. How do we regulate these dangers? Let us have a look at what we have done in the past. In 1946 the "father of the atomic bomb," Robert Oppenheimer, and Franklin Roosevelt's science adviser, Vannevar Bush, served on a government committee on nuclear arms control. The resulting report, commonly known as the Acheson-Lilienthal Plan, proposed that nuclear warfare and nuclear weapons should not be controlled by regulating the knowledge and science required to make an atomic bomb, but instead the distribution of enriched fissile materials should be tightly controlled. The plan was never accepted by either the United States or the Soviet Union, and instead the proliferation and use of nuclear weapons has been controlled by a fear of mutual destruction. This has not prevented other countries from developing nuclear weapons. The threat of someone using a nuclear bomb has been hanging over us for more than 70 years, yet we have been unable to find a way of regulating the production of atomic bombs. This despite the fact that atomic bombs have no utility; they serve no purpose other than to act as a threat, a deterrent, and ultimately a weapon of great destruction. Furthermore, building an atomic bomb is difficult to do stealthily, as enriching nuclear material requires a significant infrastructure and many workers with advanced technological and scientific skills. How are we going to regulate the weaponization of CRISPR gene drives when CRISPR and gene drives are used in labs all over the world, and the production of supercharged viruses and bacteria will be relatively cheap and require a minimal infrastructure? Perhaps this is a case where we need to limit both the practice of science and the distribution of scientific knowledge.

This isn't the first time that science has reached the point where prudence has suggested we apply the brakes. On at least two occasions, large groups of scientists have voluntarily agreed to stop their research so that ethical and safety procedures could be established. In 1974, molecular biologists voluntarily stopped doing genetic experiments out of concern for the potential consequences of their work. The following year, 180 molecular biologists from all over the world came together in Asilomar, California, to draft guidelines for safe genetic engineering that were adopted by funding agencies and became law in many countries, including the United States. But they have never been updated and are no longer sufficient to deal with modern gene-editing techniques.

The second case is related to the swine flu. Generally, as was the case in the 2009 swine flu outbreak, the swine virus is incredibly efficient at moving from person to person, but its fatality rate is low. Many people get sick but not very sick. In contrast, the avian virus is lethal when it infects humans, but fortunately it does not spread very well. Virologists and epidemiologists fear the natural formation of hybrid viruses that combine the properties of the deadly avian virus with the high transmissibility of the swine virus.

In 2011, researchers reported having successfully genetically altered the bird flu virus so it could rapidly spread among ferrets, a model organism for people. They were trying to understand how the avian flu was transmitted and to find out which mutations would be transmissible among humans, as this would help in monitoring for new mutant bird flu viruses appearing in the wild that could spread among humans. In early 2012, the danger to the public and the public concern arising from this research caused the 40 leading experts in the field to stop their work for an entire year while satisfactory safety guidelines were devised. They restarted their research in January 2013. At the same time, the U.S. government imposed a three-year funding moratorium on research into ways that viruses can be made more virulent. On December 19, 2017, the U.S. Department of Health and Human Services lifted the moratorium.

When ethical dilemmas like the sudden ability to genetically modify human embryos arise, there are no standard procedures in place to deal with them. Ad hoc solutions are found instead, such as the international committee convened by the U.S. National Academy of Sciences and the National

Academy of Medicine in Washington, DC, which in February 2017 concluded that human embryo editing may only be permitted "for compelling reasons and under strict oversight."

To date, all global science regulations have come from scientists themselves, and they have been reactive, responding to advances that were too dangerous to ignore. However, the time has come to introduce a trusted, nonpartisan advisory group of international scientists who can proactively devise safety and ethical guidelines that can be used by scientists, legislators, and funding agencies across the world. There are organizations such as the World Commission on the Ethics of Scientific Knowledge and Technology, which was set up by the United Nations Educational, Scientific and Cultural Organization, that could be the permanent solution, but they lack the necessary prominence, resources, and legitimacy.

A very interesting model scientists could use is the International Accounting Standards Foundation, which is an independent, not-for-profit organization. Its primary mission is to develop, in the public interest, a single set of high-quality, understandable, enforceable, and globally accepted accounting standards. The foundation has a three-tier structure. There is a board of trustees that is responsible for the governance and oversight of the standards, a monitoring board that facilitates interactions between the foundation and the public authorities, and a board that is responsible for setting the standards themselves. The standards board is made up of 14 experts, each with one vote. They are selected from diverse geographic backgrounds and have a mix of experience in standard-setting, preparing and using accounts, and academic work. The board uses a thorough, transparent, and participatory process to set its standards.

An International Science Standards Foundation Board based on this model would proactively set ethical and safety guidelines. Governments, journals, funders, industry, and academia would require compliance with the standards and be part of the monitoring board that interacts with the standards board. By refusing to patent, fund, tenure, or publish any research that did not follow the guidelines, the foundation might be able to prevent competition, greed, and ambition from driving science into dangerous and immoral areas. A branch of the foundation could even be responsible for providing a stamp of authenticity to journals and medical procedures to help distinguish them from predatory journals and quacks.

George Mallory, who famously justified his attempts to climb Mount Everest with the quote "because it's there" in a 1923 *New York Times* interview, would be shocked to see the lines waiting to go up the Hillary Step of Mt. Everest. He could never have imagined that in less than a century the fierce, unconquerable peak that took his life would become the playground of the rich, a natural wonder with no trash cans. What will our urge to understand all of science bring us? Where will human genetic modifications take us? Just because it's there doesn't mean we have to explore all of science; perhaps some mountains are best left unclimbed.

CONCLUSION: THE FUTURE OF SCIENCE IS BRIGHT

In my opinion, science itself is healthy. Nature's secrets are steadily being unpacked, our understanding of the universe is expanding, and we are continuing to utilize our knowledge. Earlier chapters covered the potential of deep learning, the Laser Interferometer Gravitational-Wave Observatory, CRISPR, and optogenetics. New technologies resulting from these and other new scientific developments will change our lifestyles and lengthen our lives. In order to be sustainable—at the same time that we admire the strides of modern science we must try to curb its excesses—we have to continue to support basic science, as it will form the seeds of new breakthroughs. Somewhere out there are researchers working on some bland, innocuous projects that will be the future equivalents of finding the palindromic repeats central to CRISPR, the study of jellyfish bioluminescence that led to fluorescent proteins, and the research into algae light-spots that birthed optogenetics.

But, and this is a big but, I think the future of science is almost *too* bright. Our scientific abilities and knowledge are increasing at a faster and faster rate, and as a consequence science is outgrowing its supporting structures. Governments, funding agencies, philosophers, and scientists themselves are struggling to find ways to make the most of the power of science, CRISPR, and deep learning while keeping potential abuses at bay. With increasing scientific ability the question "Are we being good ancestors?," posed by Dr. Salk, becomes even more relevant. We are in charge of our own evolution and the health of the planet; that's a lot of responsibility.

If we can find a way to appropriately regulate future discoveries, communicate with the public effectively, and continue to hold ourselves accountable, then science has nothing to worry about, but only time will tell if we'll be able to accomplish all this.

Acknowledgments

Good science, bad science, new science, old science
Applied science, basic science, wow science, now science
Say, what a lot of science there is

Many friends helped cook the book
They shook out the gobbledygook
They read the infant book

In reads of my many many words
They found science seeds among the science weeds
Their reads were selfless deeds
They squashed many false leads,
Grammatical misdeeds
And factual stampedes

Brick by brick, word by fact, I wrote *The State of Science*
Say, what a lot of help I got
One thanks, two thanks, monomeric thanks, polymeric thanks

This is a unique opportunity to not only thank everyone who contributed to the writing of this book, but also to thank those who molded me into the type of person who would enjoy writing a book about science.

THE ROAD TO SCIENCE

There are many roads to science. Mine didn't pass through chemistry sets, electronics, or telescopes. I grew up in South Africa. In our various VW Beetles, my parents took me to many nature sanctuaries and game reserves.

Through secretary birds, marabou storks, and honey badgers I was sucked into the world of nature. There were no white lab coats in my dreams. In my fantasies my skinny legs rattled around in a khaki game warden's uniform.

Mrs. Munting was my high school biology teacher. At Sasolburg High School, teachers were a different genus and species to students, and I have no clue what her first name was. She was "Mrs. Munting," and she made biology real. She took our tattered standard textbooks and blew life into them. I can't remember my chemistry teacher's name. He was a villainous sadist who would be worthy of all the derision and scorn that could be heaped upon him in a whole series of Roald Dahl books. It took Professors Rob Hancock, Arthur Howard, Wally Orchard, Neil Coville, Gus Gerans, and Jo Michael, a slew of inspiring and dedicated chemistry professors at the University of Witwatersrand, and an unfortunate experience with a botany class (involving a well-deserved F; it helps to attend classes) to divert me from a life of protecting rhinos to one of thinking in molecules.

Professors Rob Hancock (Wits), Nik Kildahl (W.P.I.), Bob Crabtree (Yale), Bruce Branchini and Stan Ching (Connecticut College), and Peter Comba (Heidelberg University) took me to the next level. They taught me how to do my own research, to write papers and grants. Thanks to them I gained an appreciation and enjoyment of science.

Similarly limited squash abilities and tastes in music, books, and humor drew me into the orbit of Gene Gallagher, a professor of religious studies at Connecticut College. Gene taught with his whole mind, body, and soul. Thanks to Gene, I have come to the realization that teaching and doing research with college-aged students is something to be taken seriously and that it is a joy and privilege.

That said, I have to thank generations of my students for keeping me young and in tune with what is happening in our world.

THE STATE OF SCIENCE: GOOD SCIENCE, BAD SCIENCE, NEW SCIENCE, OLD SCIENCE

Stuart Vyse, author of *Going Broke: Why Americans (Still) Can't Hold on to Their Money* and *Believing in Magic: The Psychology of Superstition* (both published by Oxford University Press), and Professor Tanya Schneider (Connecticut College) have read and constructively critiqued my work and provided support as I worked my way through the initial book proposal and large swaths of the manuscript. Professors Zofia Bauman (University of Connecticut, Avery Point), Jacob Stewart (Connecticut College), and Katie Launer Felty (Connecticut College) read chapters relevant to their expertise and helped me minimize the factual errors in those chapters; all the remaining errors are on me. Kayla Kibbe (Connecticut College '18) and Elizabeth

Sisko (Holy Cross College '19) read through the whole manuscript and provided valuable input. Elizabeth Berry, Ann Monk, and Medha Parnas (Connecticut College '21) worked on the introductory chapter as part of our Perspectives on Modern Society course.

PUBLISHED IDEAS, PUBLIC IDEAS, TALK IDEAS, LONG WALK IDEAS

I am teaching reactions and stoichiometry in my introductory chemistry class this week, so it should not be surprising that I am seeing *The State of Science* as a reaction. The reactants were peer-reviewed papers, books, magazine and newspaper articles, podcasts, interviews, and talks with colleagues interspersed with a sprinkling of my own thoughts. I have to thank Professors Susan Bourne (University of Cape Town), Sarah Reisman (Caltech), Maryam Foroozesh (Xavier University, New Orleans), Mehnaaz Ali (Xavier University, New Orleans), Kathleen Morgan (Xavier University, New Orleans), Tanya Schneider (Connecticut College), John McKnight (Connecticut College), and Ellen Jorgensen (Biotech without Borders) for sharing their thoughts and expertise with me. Thoughts and ideas loaned from Yuval Noah Harari, Bill McKibben, Jennifer Doudna, John Oliver, and Kai-Fu Lee are found throughout the book. And I can't forget to thank Phil and Linda Lader for introducing me to so many new scientists and new ideas.

The external reviewers of the initial proposal for the book gave me vital feedback and were crucial in helping me find a direction for the book. My agent, Jessica Papin, was the catalyst for its production. Steven Mitchell acquired the book for Prometheus Press just before retiring, and Jon Kurtz edited the book.

THE STATE OF ME: GOOD MARC, BAD MARC, HAPPY MARC, SAD MARC

It is impossible for me to write any type of acknowledgments without thanking my parents and my family (Dianne, Caitlin, Matthew, and Yuxing). They have ensured that there is a ton more happy Marc than there is sad Marc.

Notes

1. THE BIG PICTURE

1. Gershenfeld, N. (2018). "Ansatz," in *This idea is brilliant: Lost, overlooked, and underappreciated scientific concepts everyone should know* (ed. J. Brockman), New York: HarperPerennial.

2. Kelly, E. (2017). Mary Somerville: The woman for whom the word "scientist" was made, *AllThatsInteresting*, https://allthatsinteresting.com/mary-somerville.

3. Secord, J. (2018). Mary Somerville's vision of science, *Physics Today*, 47.

4. Bar-On, Y. M., Phillips, R., and Milo, R. (2018). The biomass distribution on Earth, *Proceedings of the National Academy of Sciences 125*(25), 6506–6511.

5. Harari, Y. N. (2017). *Homo deus: A brief history of tomorrow*, first U.S. ed., New York: HarperCollins.

6. Subramanian, M. (2019). Anthropocene now: Influential panel votes to recognize earth's new epoch. Atomic age would mark the start of the current geologic time unit, if proposal receives final approval, *Nature News*, May 21.

7. McKibben, B. (2019). *Falter: Has the human game begun to play itself out?* New York: Henry Holt.

8. Ball, P. (2018). "Demographics," in *What the future looks like: Scientists predict the next great discoveries and reveal how today's breakthroughs are already shaping our world* (ed. J. Al-Khalili), New York: The Experiment.

9. Harari, Y. N. *Homo deus*.

10. Harari, Y. N. (2018). *21 lessons for the 21st century*, first ed., New York: Spiegel & Grau.

11. Giussani, B. (2018). "Exponential," in *This idea is brilliant: Lost, overlooked, and underappreciated scientific concepts everyone should know* (ed. J. Brockman), New York: HarperPerennial.

12. Quoted in Yong, E. (2018). A controversial virus study reveals a critical flaw in how science is done: After researchers resurrected a long-dead pox, some critics argue that it's too easy for scientists to make decisions of global consequence, *The Atlantic*, October 4.

13. Yong, E. A controversial virus study.

14. Popovich, N., Albeck-Ripka, L., and Pierre-Louis, K. (2019). 83 environmental rules being rolled back under Trump, *New York Times*, June 7.

15. Sullivan, M., and Sellers, C. (2019). The EPA has backed off enforcement under Trump—here are the numbers, *The Conversation*, January 3.

16. Environmental Protection Agency. (2019). Our nation's air—Air quality improves as America grows, July 1, https://www.epa.gov/sites/production/files/2019-2007/documents/air trendsreport_07012019_custom_v07012011_d07012011.pdf.

17. Schreckinger, B. (2018). Trump acknowledges climate change—At his golf course, *Politico,* May 23, https://www.politico.com/story/2016/05/donald-trump-climate-change-golf-course-223436.

18. Funk, C., and Rainie, L. (2015). Public and scientists' views on science and society, Pew Research Center, January 29.

19. Crease, R. (2019). *The workshop and the world: What ten thinkers can teach us about science and authority,* New York: W. W. Norton & Company.

20. Mahler, J. (2016). The problem with "self-investigation" in a post-truth era, *The New York Times Magazine,* December 27.

21. Otto, S. L. (2016). *The war on science: Who's waging it, why it matters, what we can do about it,* first ed., Minneapolis: Milkweed Editions.

2. THE PROFESSIONAL SCIENTIST

1. McBride, J. (2018). Nobel laureate Donna Strickland: "I see myself as a scientist, not a woman in science," *The Guardian,* October 20.

2. Institute of Education Sciences—National Center for Education Statistics. (2016). Total undergraduate fall enrollment in degree-granting postsecondary institutions, by attendance status, sex of student, and control and level of institution: Selected years, 1970 through 2026, https://nces.ed.gov/programs/digest/d16/tables/dt16_303.70.asp.

3. National Center for Education Statistics. (2016). Table 318.45: Number and percentage distribution of science, technology, engineering, and mathematics (STEM) degrees/certificates conferred by postsecondary institutions, by race/ethnicity, level of degree/certificate, and sex of student: 2008–2009 through 2014–2015, *Digest of education statistics: 2016 tables and figures,* https://nces.ed.gov/programs/digest/d16/tables/dt16_318.45.asp.

4. Institute of Education Sciences—National Center for Education Statistics. (2016). Total undergraduate fall enrollment in degree-granting postsecondary institutions, by attendance status, sex of student, and control and level of institution: Selected years, 1970 through 2026, https://nces.ed.gov/programs/digest/d16/tables/dt16_303.70.asp.

5. National Science Foundation. (2017). Women, minorities, and persons with disabilities in science and engineering, https://www.nsf.gov/statistics/2017/nsf17310/static/downloads/ nsf17310-digest.pdf; Ashanti Johnson, M. H. O. (2016). How to recruit and retain underrepresented minorities, *American Scientist 104,* 76–91.

6. National Center for Science and Engineering Statistics. (2017). Women, minorities, and persons with disabilities in science and engineering, https://www.nsf.gov/news/ news_summ.jsp?cntn_id=190946.

7. Harris, A. (2019). The disciplines where no black people earn Ph.Ds: In more than a dozen academic fields—largely STEM related—not a single black student earned a doctoral degree in 2017, *The Atlantic,* April 19.

8. James, R., Starks, H., Segrest, V. A., and Burke, W. (2012). From leaky pipeline to irrigation system: Minority education through the lens of community-based participatory research, *Progress in Community Health Partnerships: Research, Education, and Action 6,* 471–479.

9. Pennington, C. R., Heim, D., Levy, A. R., and Larkin, D. T. (2016). Twenty years of stereotype threat research: A review of psychological mediators, *PLOS One 11,* e0146487–e0146487.

10. Yosso, T., Smith, W., Ceja, M., and Solórzano, D. (2009). Critical race theory, racial microaggressions, and campus racial climate for Latina/o undergraduates, *Harvard Educational Review 79,* 659–691; Sue, D. W., Capodilupo, C. M., Torino, G. C., Bucceri, J. M., Holder, A. M. B., Nadal, K. L., and Esquilin, M. (2007). Racial microaggressions in everyday life: Implications for clinical practice, *American Psychology 62,* 271–286.

11. Harris, A. The disciplines where no black people earn Ph.Ds.

12. Xavier University of Louisiana, Office of Planning, Institutional Research and Assessment. (2018–2019). University profile.

13. *Diverse: Issues in Higher Education*. (2019). Top 100 degree producers, https://diverse education.com/top100/pages/index.php.

14. National Science Foundation. (2013, April). Baccalaureate origins of U.S.-trained science and engineering doctorate recipients, https://www.nsf.gov/statistics/infbrief/nsf13323/.

15. Association of American Medical Colleges. (2012). Diversity facts and figures, https://www.aamc.org/data-reports/workforce/report/diversity-facts-figures.

16. Mervis, J. (2014). When it comes to diversity grants, NIH hopes bigger is better, *Science*, November 4.

17. Morgan, W. (2018). No black scientist has ever won a Nobel—that's bad for science, and bad for society, *The Conversation*, October 8.

18. Board, N. S. (2018). Science and engineering indicators 2018, National Science Foundation, https://www.nsf.gov/statistics/2018/nsb20181/.

19. Redden, E. (2018). International student numbers decline, *Inside Higher Education*, January 22.

20. Redden, E. (2018). New international enrollments decline again, *Inside Higher Education*, November 13.

21. Royal Swedish Academy of Sciences. (2018). Press release: The 2018 Nobel Prize in Physics, https://www.nobelprize.org/prizes/physics/2018/press-release/.

22. McBride, J. Nobel laureate Donna Strickland.

23. Paul, A. (2018). Five women who missed out on the Nobel prize, *The Guardian*, October 7.

24. Gibney, E. (2018). Nobel laureate Donna Strickland talks lasers and gender, *Nature Podcast*, https://www.nature.com/articles/d41586-018-06995-w.

25. Ford, K. (2017). Defeating the inner imposter that keeps us from being successful, *TEDxMidAtlantic*, February 22.

26. Valian, V. (2018). Two Nobels for women—Why so slow?, *Nature*, https://www.nature.com/articles/d41586-018-06953-6.

27. The Nobel Prize. (2018). Nomination and selection of Nobel laureates, https://www.nobelprize.org/nomination-and-selection-of-nobel-laureates/.

28. Garbee, E. (2017). The problem with the "pipeline": A pervasive metaphor in STEM education has some serious flaws, *Slate*, October 17.

29. Garbee, E. The problem with the "pipeline."

30. Maxman, A. (2018). Why it's hard to prove gender discrimination in science, *Nature News & Comment*, May 15.

31. Jarvis, L. M. (2018). Why can't the drug industry solve its gender diversity problem?, *Chemistry & Engineering News*, 96.

32. Urry, M. (2015). Science and gender: Scientists must work harder on equality, *Nature-Comments*, December 21.

33. MassBio. (2017). A study of gender diversity within the life sciences sector of Massachusetts, http://files.massbio.org/file/MassBio-Liftstream-Gender-Diversity-Report-2017-C849.PDF.

34. Fara, P. (2018). "Leaky pipelines": Plug the holes or change the system?, *National Public Radio: Cosmos and Culture*, February 2, 2018.

35. Mason, M. A., and Goulden, M. (2002). Do babies matter? The effect of family formation on the lifelong careers of academic men and women, *Academe 88*, 21–27.

36. Trix, F., and Psenka, C. (2003). Exploring the color of glass: Letters of recommendation for female and male medical faculty, *Discourse & Society 14*, 191–220.

37. Knobloch-Westerwick, S., Glynn, C. J., and Huge, M. (2013). The Matilda effect in science communication: An experiment on gender bias in publication quality perceptions and collaboration interest, *Science Communication 35*, 603–625.

38. Moss-Racusin, C. A., Dovidio, J. F., Brescoll, V. L., Graham, M. J., and Handelsman, J. (2012). Science faculty's subtle gender biases favor male students, *Proceedings of the National Academy of Sciences 109*, 16474.

39. Valian, V. (1998). *Why so slow? The advancement of women*, Cambridge, MA: MIT Press.

40. Nittrouer, C. L., Hebl, M. R., Ashburn-Nardo, L., Trump-Steele, R. C. E., Lane, D. M., and Valian, V. (2018). Gender disparities in colloquium speakers at top universities, *Proceedings of the National Academy of Sciences 115*, 104.

41. Nittrouer et al. Gender disparities in colloquium speakers at top universities.

42. Quoted in Maxman, A. (2018). Why it's hard to prove gender discrimination in science, *Nature News*, May 15.

43. Quoted in Maxman, A. Why it's hard to prove gender discrimination in science.

44. Funk, C., and Parker, K. (2018). Women and men in STEM often at odds over workplace equity, Pew Research Center, January 9.

45. Goodwin, K. (2018). Mansplaining, explained in one simple chart, BBC, July 29.

46. Cooper, K. M., Krieg, A., and Brownell, S. E. (2018). Who perceives they are smarter? Exploring the influence of student characteristics on student academic self-concept in physiology, *Advances in Physiology Education 42*, 200–208.

47. Fox, M. (2018). Not smart enough? Men overestimate intelligence in science class: Even when grades show different, men overestimated their class ranking, NBC News, April 4.

48. Fox, M. Not smart enough?

49. Fine, I., and Shen, A. (2018). Perish not publish? New study quantifies the lack of female authors in scientific journals, *The Conversation*, March 8.

50. Fine, I., and Shen, A. Perish Not Publish?

51. Macaluso, B., Lariviere, V., Sugimoto, T., and Sugimoto, C. R. (2016). Is science built on the shoulders of women? A study of gender differences in contributorship, *Academic Medicine 91*, 1136–1142.

52. Storage, D., Horne, Z., Cimpian, A., and Leslie, S. J. (2016). The frequency of "brilliant" and "genius" in teaching evaluations predicts the representation of women and African Americans across fields, *PLOS One 11*, e0150194.

53. Fine, I., and Shen, A. Perish not publish?

54. Staley, O., and Shendruk, A. (2018). Here's what the stark gender disparity among top orchestra musicians looks like, *Quartz at Work*, October 16.

55. Carter, A. J., Croft, A., Lukas, D., and Sandstrom, G. M. (2018). Women's visibility in academic seminars: Women ask fewer questions than men, *PLOS One 8*, e0212146.

56. Jarvis, L. M. Why can't the drug industry solve its gender diversity problem?

57. Bear, J. B., and Woolley, A. W. (2011). The role of gender in team collaboration and performance, *Interdisciplinary Science Reviews 36*, 146–153; Smith, Wendy K. T., and Babcock-Lumish, T. (2018). How many women does it take to change a broken Congress?, *The Conversation*, November 9.

58. Koren, M. (2018). One Wikipedia page is a metaphor for the Nobel Prize's record with women, *The Atlantic*, October 2.

59. Lander, E. S. (2018). Will America yield its position as the world's leader in science and technology?, *Boston Globe*, January 29.

60. Harari, Y. N. (2014). *Sapiens: A brief history of humankind*, London: Harvill Secker.

61. Gracia, S. (2018). Nobel Prize in chemistry goes to a woman for the fifth time in history, *New York Times*, October 3.

3. DO-IT-YOURSELF SCIENCE

1. Hannibal, M. E. (2016). *Citizen scientist: Searching for heroes and hope in an age of extinction*, New York: The Experiment; Hannibal, M. E. (2017). Can amateur scientists save animals from extinction?, *NPR TED Radio Hour*, September 29.

2. Sacks, O. (1996). *An anthropologist on Mars: Seven paradoxical tales*, New York: Vintage Books.

3. Bryson, B. (2003). *A short history of nearly everything,* New York: Broadway Books.

4. Koshland, D. E. (1992). Where the grass is rougher and greener, *Science 257*, 1607.

5. Berman, R. (2018). It's a movement: Amateur scientists are making huge discoveries: Citizen scientists are advancing scientific knowledge, *Big Think*, February 3.

6. Riesch, H., and Potter, C. (2014). Citizen science as seen by scientists: Methodological, epistemological and ethical dimensions, *Public Understanding of Science 23*, 107.

7. Quoted in Oberhaus, D. (2015). Seven ways to donate your computer's unused processing power: Help cure cancer and find aliens while you sleep, *Motherboard*, September 7.

8. Quoted in Oberhaus, D. Seven ways to donate your computer's unused processing power.

9. Matchar, E. (2017). AI plant and animal identification helps us all be citizen scientists: Apps that use artificial intelligence to allow users to ID unknown specimens are making science more accessible to everyone, *Smithsonian.com*, June 7.

10. Global Biodiversity Information Facility. (2018). What is GBIF?, https://www.gbif.org/what-is-gbif.

11. Irwin, A. Citizen science comes of age, *Nature 462* (October 2018).

12. Zimmer, M. (2005). *Glowing genes: A revolution in biotechnology*, Amherst, NY: Prometheus Books.

13. Zimmer, M. *Glowing genes*.

14. Schouweiler, S. (2009). Art world today will meet "Edunia," Eduardo Kac's genetically engineered "plantimal," *Minnpost*, April 17.

15. Jorgensen, E. (2012). Biohacking—You can do it, too, *TEDGlobal*, https://www.ted.com/talks/ellen_jorgensen_biohacking_you_can_do_it_too?language=en.

16. Grushkin, D. (2018). Biohackers are about open access to science, not DIY pandemics: Stop misrepresenting us, *STATNews*, June 4.

17. Lipinski, J. (2011). On Flatbush Avenue, seven stories full of ideas, *New York Times*, January 11.

18. Jorgensen, E. Biohacking.

19. Grushkin, D., Kuiken, T., and Millet, P. (2013). Seven myths and realities about do-it-yourself biology, Woodrow Wilson Center.

20. Grushkin, Kuiken, and Millet. Seven myths and realities about do-it-yourself biology.

21. Darnovsky, M. (2018). Hacking your own fenes: A recipe for disaster, *Leapsmag*, January 17.

22. Zayner, J. (2017). How to genetically engineer a human in your garage, YouTube, February 15, https://www.youtube.com/watch?v=imTXcEh79lw.

23. Darnovsky, M. Hacking your own genes.

24. Quoted in Brown, K. V. (2017). Genetically engineering yourself sounds like a horrible idea—But this guy is doing it anyway, *Gizmodo*, November 29.

25. Regalado, A. (2019). Don't change your DNA at home, says America's first CRISPR law: A California "human biohacking" bill calls for warnings on do-it-yourself genetic-engineering kits, *MIT Technology Review*, August 9, https://www.technologyreview.com/s/614100/dont-change-your-dna-at-home-says-americas-first-crispr-law/.

26. Zayner, J. How to genetically engineer a human in your garage.

27. Brown, K. V. Genetically engineering yourself sounds like a horrible idea.

28. Quoted in Griffen, P. (2018). Edit thyself: Biohacking in the age of CRISPR, *Science in the News*, Harvard University, The Graduate School of Arts and Sciences, http://sitn.hms.harvard.edu/flash/2018/edit-thyself-biohacking-age-crispr/.

4. THE NUTS AND BOLTS

1. Somers, J. (2018). The scientific paper is obsolete, *The Atlantic*, April 5.

2. Zimmer, M. (2007). Guerrilla puzzling: A model for research, *Chronicle of Higher Education*, February, B5.

3. Ware, M., and Mabe, M. (2009). The STM report—An overview of scientific and scholarly journals publishing, International Association of Scientific, Technical and Medical Publishers, https://www.stm-assoc.org/2012_12_11_STM_Report_2012.pdf.

4. *Nature.* (2018). Editorial policies, https://www.nature.com/nature/for-authors/editorial-criteria-and-processes.

5. Carroll, A. E. (2018). Peer review: The worst way to judge research, except for all the others—A look at the system's weaknesses, and possible ways to combat them, *New York Times*, November 5.

6. Vesper, I. (2018). Peer reviewers unmasked: Largest global survey reveals trends, *Nature*, https://www.nature.com/articles/d41586-018-06602-y, doi:10.1038/d41586-41018-06602-y; Publons. (2018). Global state of peer review: 2018, https://publons.com/static/Publons-Global-State-Of-Peer-Review-2018.pdf.

7. Carroll, A. E. Peer review.

8. Chuang, K. V., Xu, C., and Reisman, S. E. (2016). A 15-step synthesis of (+)-ryanodol, *Science 353*, 912–915.

9. Quoted in Halford, B. (2016). Rapid route to ryanodol, *Chemical & Engineering News*, ACS meeting coverage, http://acsmeetings.cenmag.org/rapid-route-to-ryanodol/.

10. Powell, K. (2016). The waiting game: Scientists are becoming increasingly frustrated by how long it seems to take to publish papers, but is it really getting worse?, *Nature 530*, 148.

11. Quoted in Powell, K. The waiting game.

12. Bornmann, L., and Mutz, R. (2015). Growth rates of modern science: A bibliometric analysis based on the number of publications and cited references, *Journal of the Association for Information Science and Technology 66*, 2215–2222.

13. Quoted in Powell, K., The waiting game.

14. Vale, R. D. (2015). Accelerating scientific publication in biology, *Proceedings of the National Academy of Sciences 112*, 13439.

15. Quoted in Heaven, D. (2018). AI peer reviewers unleashed to ease publishing grind: A suite of automated tools is now available to assist with peer review but humans are still in the driver's seat, *Nature News*, November 22.

16. Van Noorden, R., Maher, B., and Nuzzo, R. (2014). "The top 100 papers: *Nature* explores the most-cited research of all time, *Nature 541*, 550.

17. Buranyi, S. (2017). Is the staggeringly profitable business of scientific publishing bad for science? It is an industry like no other, with profit margins to rival Google and it was created by one of Britain's most notorious tycoons: Robert Maxwell, *The Guardian*, June 17.

18. Publons. Global state of peer review.

19. Prokop, A. (2018). The new BSDB Newsletter: A focus on communication and advocacy, *Newsletter British Society for Developmental Biology*, January 26.

20. Harris, R. F. (2017). *Rigor mortis: How sloppy science creates worthless cures, crushes hope, and wastes billions.* Basic Books.

21. Quoted in Begley, S. (2012). In cancer science, many "discoveries" don't hold up, Reuters, March 28.

22. Edwards, M. A., and Roy, S. (2017). Academic research in the 21st century: Maintaining scientific integrity in a climate of perverse incentives and hypercompetition, *Environmental Engineering Science 34*, 51–61.

23. Finkel, A. (2018). Science isn't broken, but we can do better: Here's how, *The Conversation*, April 17.

24. Ioannidis, J. P. A., Klavans, R., and Boyack, K. W. (2018). Thousands of scientists publish a paper every five days, *Nature 561*, 167–169.

25. Stern, B. M., and O'Shea, E. K. (2018). Scientific publishing in the digital age, *ASAPbio*, January 30.

26. Quoted in Van Noorden, R. (2013). Open access: The true cost of science publishing, *Nature 495*, 426–429.

27. Quoted in Kwon, D. (2018). Plan S: The ambitious initiative to end the reign of paywalls: A funder-driven push for freely accessible scholarly literature has divided the scientific community, *The Scientist*, December 19.

28. Kaiser, J. (2017). Are preprints the future of biology? A survival guide for scientists, *Science*, September 29.

29. Quoted in Powell, K. The waiting game.

30. Graber-Stiehl, I. (2018). Meet the pirate queen making academic papers free online: Science's pirate queen—Alexandra Elbakyan is plundering the academic publishing establishment, *The Verge*, February 8.

31. Kirchherr, J. (2018). A PhD should be about improving society, not chasing academic kudos, *The Guardian*, August 9.

32. Kirchherr, J. A PhD should be about improving society, not chasing academic kudos.

33. Mervis, J. (2017). Data check: U.S. government share of basic research funding falls below 50%, *Science*, March 9.

34. Sinkjær, T. (2018). Fund ideas, not pedigree, to find fresh insight, *Nature 555*, 143.

35. Leeming, J. (2018). How researchers are ensuring that their work has an impact: Finding purpose and meaning in the lab, *Nature 556*, 139–141.

36. Patel, N. V. (2015). Is crowdfunded science crap?, *Slate*, April 27; Patel, N. V. (2018). How interesting does science have to be for the public to fund it?, *Slate*, January 4.

37. Peters, J. (2018). Billionaires are rushing into biotech: Inequality is following them into science; "Free-market philanthropy" raises yet more questions about the future of American public research, *Massive*, May 5.

5. RECOGNIZING A BREAKTHROUGH IN SCIENCE

1. Watson, J. D., and Crick, F. H. C. (1953). Molecular structure of nucleic acids: A structure for deoxyribose nucleic acid, *Nature 171*, 737.

2. Chargaff, E., Zamenhof, S., and Green, C. (1950). Human desoxypentose nucleic acid: Composition of human desoxypentose nucleic acid, *Nature 165*, 756.

3. Maddox, B. (2003). The double helix and the wronged heroine, *Nature 421*, 407.

4. Watson, J. D. (1981). *The double helix: A personal account of the discovery of the structure of DNA*, a new critical edition including text, commentary, reviews, original papers, edited by Gunther S. Stent. London: Weidenfeld and Nicolson.

5. Crick, F. H. C., and Watson, J. D. (1952). The complementary structure of deoxyribonucleic acid, *Proceedings of the Royal Society A 223*, 80–96.

6. Grady, D. (2003). A revolution at 50; 50 years later, Rosalind Franklin's X-ray fuels debate, *New York Times*, February 25; Maddox, B. The double helix and the wronged heroine; Wade, N. (2003). Was she, or wasn't she?, *The Scientist*, April 7.

7. Pinker, S. (2018). *Enlightenment now: The case for reason, science, humanism, and progress*, Viking.

8. Prasher, D. C., Eckenrode, V. K., Ward, W. W., Prendergast, F. G., and Cormier, M. J. (1992). Primary structure of the Aequorea victoria green-fluorescent protein, *Gene 111*, 229–233.

9. Quoted in Talan, J. (2008). Why a little green signaling protein prompted this year's Nobel Prize in chemistry—The Nobelists share their story, *Neurology Today 8*, 15–16.

10. Nobelprize.org. (2008). Roger Y. Tsien banquet speech.

11. Quoted in Zimmer, M. (2009). GFP: From jellyfish to the Nobel Prize and beyond, *Chemical Society Reviews 38*, 2823–2832.

12. Poundstone, W. (2018). "Stigler's law of economy," in *This idea is brilliant: Lost, overlooked, and underappreciated scientific concepts everyone should know* (ed. J. Brockman), New York: HarperPerennial.

6. LIGO AND VIRGO

1. Kaiser, D. (2017). Learning from gravitational waves, *New York Times*, October 3.

2. Twilley, N. (2016). How the first gravitational waves were found: After decades of speculation and searching, a signal came through; it promises to change our understanding of the universe, *New Yorker*, February 11.

3. Quoted in Twilley, N. How the first gravitational waves were found.

4. Twilley, N. How the first gravitational waves were found.

5. Abbott, B. P., et al. (2016). Observation of gravitational waves from a binary black hole merger, *Physical Review Letters 116*, 061102.

6. Thorne, K. A. (2014). *The science of* Interstellar, W. W. Norton & Company.

7. Ward, R. L. (2018). We're going to get a better detector: Time for upgrades in the search for gravitational waves, *The Conversation*, August 16, https://theconversation.com/were-going-to-get-a-better-detector-time-for-upgrades-in-the-search-for-gravitational-waves-100382.

8. Padma, T. V. (2019). India's LIGO gravitational-wave observatory gets green light: The US$177-million observatory will join a global network of sensors and should improve sensitivity, *Nature News*, January 22, https://www.nature.com/articles/d41586-41019-00184-z.

9. Botner, O. (2017). The Nobel Prize in physics 2017, NobelPrize.org, Nobel Media, https://www.nobelprize.org/prizes/physics/2017/summary/.

10. Quoted in Overbye, D. (2016). Gravitational waves detected, confirming Einstein's theory, *New York Times*, February 11.

11. Overbye, D. Gravitational waves detected, confirming Einstein's theory.

12. Wu, L., Wang, D., and Evans, J. A. (2019). Large teams develop and small teams disrupt science and technology, *Nature 566*, 378–382.

13. Quoted in Yong, E. (2019). Small teams of scientists have fresher ideas: A new study shows that little teams are more likely to take their research in radically new directions, *The Atlantic*, February 13.

7. DEEP LEARNING

1. Lee, K.-F. (2018). *AI superpowers: China, Silicon Valley, and the new world order*, Boston: Houghton Mifflin Harcourt.

2. Topol, E. (2019). *Deep medicine: How artificial intelligence can make healthcare human again*, Basic Books.

3. Lee, K.-F. *AI superpowers*.

4. Kasparov, G. (2018). Chess, a drosophila of reasoning, *Science 362*, 1087.

5. Kasparov, G. Chess, a Drosophila of reasoning.

6. Lee, K.-F. *AI superpowers*.

7. Brown, N., and Sandholm, T. (2019). Superhuman AI for multiplayer poker, *Science*, August 30, https://science.sciencemag.org/content/early/2019/07/10/science.aay2400.

8. Fiedler, I., and Wilcke, A.-C. (2011). The market for online poker, *SSRN Electronic Journal 16*, 5.

9. Sample, I. (2018). Google's DeepMind predicts 3D shapes of proteins—AI program's understanding of proteins could usher in new era of medical progress, *The Guardian*, December 2.

10. Quoted in Sample, I. Google's DeepMind predicts 3D shapes of proteins.

11. *Transport Topics*. (2014). iTECH: Trucking may save $168 billion annually with driverless vehicles, report concludes, January 1.

12. Lee, K.-F. *AI superpowers*.

13. Topol, E. *Deep medicine*.

14. Topol, E. *Deep medicine*.

15. Grisham, S. (2017). *Medscape physician compensation report 2017*, Medscape.

16. Topol, E. *Deep medicine*.

17. Lee, K.-F. *AI superpowers*.

18. Diamandis, P. H. (2019). AI is rapidly augmenting healthcare and longevity, *Singularity Hub*, February, 15.

19. Segler, M. H. S., Preuss, M., and Waller, M. P. (2018). Planning chemical syntheses with deep neural networks and symbolic AI, *Nature 555*, 604.

20. Harari, Y. N. (2018). Why technology favors tyranny, *The Atlantic*, October.

21. Spencer, M. (2019). Artificial intelligence regulation may be impossible, *Forbes*, March 2.

8. OPTOGENETICS

1. Wheeler, J. (2016). EP16: Jesse Wheeler/DragonflEye, https://vimeo.com/187043804.

2. Han, W., Tellez, L. A., Rangel, M., Motta, S. C., Zhang, X., Perez, I. O., Canteras, N. S., Shammah-Lagnado, S. J., van den Pol, A. N., and de Araujo, I. E. (2017). Integrated control of predatory hunting by the central nucleus of the amygdala, *Cell 168*, 311–324, e318.

3. Zhou, T., Zhu, H., Fan, Z., Wang, F., Chen, Y., Liang, H., Yang, Z., Zhang, L., Lin, L., Zhan, Y., Wang, Z., and Hu, H. (2017). History of winning remodels thalamo-PFC circuit to reinforce social dominance, *Science 357*, 162.

4. Crick, F. (1999). The impact of molecular biology on neuroscience, *Philosophical Transactions of the Royal Society of London, Series B: Biological Sciences 354*, 2021.

5. Nagel, G., Szellas, T., Huhn, W., Kateriya, S., Adeishvili, N., Berthold, P., Ollig, D., Hegemann, P., and Bamberg, E. (2003). Channelrhodopsin-2, a directly light-gated cation-selective membrane channel, *Proceedings of the National Academy of Sciences 100*, 13940.

6. Boyden, E. S., Zhang, F., Bamberg, E., Nagel, G., and Deisseroth, K. (2005). Millisecond-timescale, genetically targeted optical control of neural activity, *Nature Neuroscience 8*, 1263–1268.

7. Quoted in Zhang, S. (2015). The unexpected science of manipulating neurons with light, *Wired*, September 8, http://www.wired.com/2015/09/unexpected-science-manipulating-neurons-light.

8. Zhang, F., Wang, L. P., Brauner, M., Liewald, J. F., Kay, K., Watzke, N., Wood, P. G., Bamberg, E., Nagel, G., Gottschalk, A., and Deisseroth, K. (2007). Multimodal fast optical interrogation of neural circuitry, *Nature 446*, 633–U634.

9. Deisseroth, K. (2008). Thy-1::ChR2-EYFP mouse, YouTube, https://www.youtube.com/watch?v=88TVQZUfYGw.

10. Quoted in Ridgeway, L. (2011). Gene therapy has potential to restore sight to the blind, *USC News*, April 20, https://news.usc.edu/29397/Gene-Therapy-Has-Potential-to-Restore-Sight-to-the-Blind.

11. Horsager, A., and Boyden, E. (2011). Blind mice, no longer, YouTube, https://www.youtube.com/watch?v=jY5Aynh1-cU.

12. Bourzac, K. (2016). Texas woman is the first person to undergo optogenetic therapy, *MIT Technology Review*, March 18.

13. Quoted in Starkman, R. (2012). Optogenetics: A novel technology with questions old and new, *Huffington Post*, September 25, http://www.huffingtonpost.com/ruth-starkman/optogenetics-a-new-techno_b_1700219.html.

14. Quoted in Schoonover, C. E., and Rabinowitz, A. (2011). Control desk for the neural switchboard, *New York Times*, May 16.

15. Bruegmann, T., Beiert, T., Vogt, C. C., Schrickel, J. W., and Sasse, P. (2018). Optogenetic termination of atrial fibrillation in mice, *Cardiovascular Research 114*, 713–723.

16. Hibberd, T. J., Feng, J., Luo, J., Yang, P., Samineni, V. K., Gereau, R. W., 4th., Kelley, N., Hu, H., and Spencer, N. J. (2018). Optogenetic induction of colonic motility in mice, *Gastroenterology 155*, 514–518.

17. Quoted in Cohen, B. (2018). Fruit flies have orgasms—And, apparently, they're amazing: In bizarre research project, scientists also discover why some fruit flies avoid getting drunk, *From the Grapevine*, April 20.

18. Zer-Krispil, S., Zak, H., Shao, L., Ben-Shaanan, S., Tordjman, L., Bentzur, A., Shmueli, A., and Shohat-Ophir, G. (2018). Ejaculation induced by the activation of Crz neurons is rewarding to Drosophila males, *Current Biology 28*, e1443.

19. Quoted in Yong, E. (2018). Scientists genetically engineered flies to ejaculate under red light: Their experiments confirm that sex is pleasurable, even for animals we think of as simple, *The Atlantic*, April 19.

20. Reuters. (2018). Researchers discover fruit flies love having orgasms, *New York Post*. May 9.

21. Ferreira, B. (2018). Male fruit flies love to cum, and turn to alcohol if they can't: A new study creates a red light district for flies, *Motherboard*, April 19.

22. Cohen, B. Fruit flies have orgasms.

23. Schuster, R. (2018). Fruit flies have mind-blowing orgasms, Israeli scientists prove— And if they haven't gotten laid, fruit flies demonstrate a desire to get drunk: Studying pleasure in Drosophila neurons could help us better understand and treat the mechanisms of addiction in human brains, researchers say, *Haaretz*, April 19.

24. Quoted in Calderone, J. (2014). 10 big ideas in 10 years of brain science, *Scientific American*, November 6.

25. Cashin-Garbutt, A. (2017). Advances in optogenetics: An interview with Dr. Karl Deisseroth, *Medical Life Science News*, March 2, https://www.news-medical.net/news/20170302/Advances-in-optogenetics.aspx.

9. CRISPR
(CLUSTERED REGULARLY INTERSPACED SHORT PALINDROMIC REPEATS)

1. Clapper, J. R. (2016). Worldwide threat assessment of the US intelligence community to the Senate Armed Services Committee, https://www.dni.gov/files/documents/SASC_Unclassified_2016_ATA_SFR_FINAL.pdf.

2. Gottlieb, S. (2018). Remarks by Commissioner Gottlieb to the Alliance for Regenerative Medicine's annual board meeting, May 22, Washington, DC, https://www.fda.gov/News-Events/Speeches/ucm608445.htm.

3. Doudna, J. A., and Sternberg, S. H. (2017). *A crack in creation: Gene editing and the unthinkable power to control evolution*, Boston: Houghton Mifflin Harcourt.

4. Quoted in Zimmer, C. (2015). Breakthrough DNA editor born of bacteria, *Quanta Magazine*, February 6.

5. Doudna, J. A., and Sternberg, S. H. *A crack in creation.*

6. Jinek, M., Chylinski, K., Fonfara, I., Hauer, M., Doudna, J. A., and Charpentier, E. (2012). A programmable dual-RNA–guided DNA endonuclease in adaptive bacterial immunity, *Science 337*, 816.

7. Gasiunas, G., Barrangou, R., Horvath, P., and Šikšnys, V. (2012). Cas9–crRNA ribonucleoprotein complex mediates specific DNA cleavage for adaptive immunity in bacteria, *Proceedings of the National Academy of Sciences 109*, E2579.

8. Zhang, S. (2015). The battle over genome editing gets science all wrong, *Wired*, October 4.

9. Cong, L., Ran, F. A., Cox, D., Lin, S., Barretto, R., Habib, N., Hsu, P. D., Wu, X., Jiang, W., Marraffini, L. A., and Zhang, F. (2013). Multiplex genome engineering using CRISPR/Cas systems, *Science 339*, 819.

10. Jinek, M., East, A., Cheng, A., Lin, S., Ma, E., and Doudna, J. (2013). RNA-programmed genome editing in human cells, *eLife 2*, e00471.

11. Brooks, J. (2019). Making sense of the CRISPR patent dispute between the University of California and Broad, *KQED*, February 20.

12. Zhang, S. (2018). The little-known nonprofit behind the CRISPR boom, *The Atlantic*, June 13.

13. Quoted in Addgene keeps flow of CRISPR plasmids fast and affordable. (2018). *EurekaAlert!*, press release, June 18.

14. Oliver, J. (2018). Gene editing, *Last Week Tonight with John Oliver*, July 1.

15. Doudna, J. A., and Sternberg, S. H. *A crack in creation.*

16. Molteni, M. (2018). Crispr'd food, coming soon to a supermarket near you, *Wired*, March 30.

17. Houser, K. (2018). USDA announces super-chill stance on gene-edited crops, *Futurism*, April 2.

18. Denby, C. M., Li, R. A., Vu, V. T., Costello, Z., Lin, W., Chan, L. J. G., Williams, J., Donaldson, B., Bamforth, C. W., Petzold, C. J., Scheller, H. V., Martin, H. G., and Keasling, J. D. (2018). Industrial brewing yeast engineered for the production of primary flavor determinants in hopped beer, *Nature Communications 9*, 965.

19. Miao, C., Xiao, L., Hua, K., Zou, C., Zhao, Y., Bressan, R. A., and Zhu, J.-K. (2018). Mutations in a subfamily of abscisic acid receptor genes promote rice growth and productivity, *Proceedings of the National Academy of Sciences 115*, 6058.

20. Quoted in Wallheimer, B. (2018). CRISPR-edited rice plants produce major boost in grain yield, *Phys.org*, https://phys.org/news/2018-05-crispr-edited-rice-major-boost-grain.html.

21. Quoted in Bunge, J. M., and Dockser, A. (2018). Is this tomato engineered? Inside the coming battle over gene-edited food, *Wall Street Journal*, April 15.

22. Ledford, H. (2017). Fixing the tomato: CRISPR edits correct plant-breeding snafu, *Nature News*, May 18; Soyk, S., Lemmon, Z. H., Oved, M., Fisher, J., Liberatore, K. L., Park, S. J., Goren, A., Jiang, K., Ramos, A., van der Knaap, E., Van Eck, J., Zamir, D., Eshed, Y., and Lippman, Z. B. (2017). Bypassing negative epistasis on yield in tomato imposed by a domestication gene, *Cell 169*, 1142–1155.

23. Li, Y. (2018). These CRISPR-modified crops don't count as GMOs, *The Conversation*, May 22.

24. Quoted in Skerrett, P. (2015). Experts debate: Are we playing with fire when we edit human genes?, *STAT*, November 17.

25. Doudna, J. A., and Sternberg, S. H. *A crack in creation.*

26. Zheng, Q., Lin, J., Huang, J., Zhang, H., Zhang, R., Zhang, X., Cao, C., Hambly, C., Qin, G., Yao, J., Song, R., Jia, Q., Wang, X., Li, Y., Zhang, N., Piao, Z., Ye, R., Speakman, J. R., Wang, H., Zhou, Q., Wang, Y., Jin, W., and Zhao, J. (2017). Reconstitution of UCP1: Using CRISPR/Cas9 in the white adipose tissue of pigs decreases fat deposition and improves thermogenic capacity, *Proceedings of the National Academy of Sciences 114*, E9474.

27. Kambadur, R., Sharma, M., Smith, T. P. L., and Bass, J. J. (1997). Mutations in myostatin (GDF8) in double-muscled Belgian blue and Piedmontese cattle, *Genome Research 7*, 910–915.

28. Proudfoot, C., Carlson, D. F., Huddart, R., Long, C. R., Pryor, J. H., King, T. J., Lillico, S. G., Mileham, A. J., McLaren, D. G., Whitelaw, C. B. A., and Fahrenkrug, S. C. (2015). Genome edited sheep and cattle, *Transgenic Research 24*, 147–153.

29. United Network for Organ Sharing. (2018). Data, June 27, https://unos.org/data/.

30. Yang, L., Güell, M., Niu, D., George, H., Lesha, E., Grishin, D., Aach, J., Shrock, E., Xu, W., Poci, J., Cortazio, R., Wilkinson, R. A., Fishman, J. A., and Church, G. (2015). Genome-wide inactivation of porcine endogenous retroviruses (PERVs), *Science 350*, 1101.

31. Niu, D., Wei, H.-J., Lin, L., George, H., Wang, T., Lee, I. H., Zhao, H.-Y., Wang, Y., Kan, Y., Shrock, E., Lesha, E., Wang, G., Luo, Y., Qing, Y., Jiao, D., Zhao, H., Zhou, X., Wang, S., Wei, H., Güell, M., Church, G. M., and Yang, L. (2017). Inactivation of porcine endogenous retrovirus in pigs using CRISPR-Cas9, *Science 357*, 1303.

32. Quoted in Perry, C. (2018). Pig organs for human patients: A challenge fit for CRISPR, *Harvard Gazette*, May 30, https://news.harvard.edu/gazette/story/2018/2005/pig-organs-for-human-patients-a-challenge-fit-for-crispr/.

33. Marcus, A. D. (2018). Meet the scientists bringing extinct species back from the dead, *Wall Street Journal*, October 11; Shapiro, B. (2016). *How to clone a mammoth: The science of de-extinction*, Princeton University Press.

34. Biba, E. (2019). Scientists could soon resurrect the woolly mammoth—But should they?, *Gizmodo*, May 11.

35. Gottlieb, S. Remarks by Commissioner Gottlieb to the Alliance for Regenerative Medicine's annual board meeting.

36. MIT (2017). Existing gene therapy pipeline likely to yield dozens of approved products within five years, New Drug Development Paradigms Initiative, Research Brief, https://new-digs.mit.edu/sites/default/files/FoCUS_Research_Brief_2017F211v011.pdf.

37. Mullin, E. (2018). CRISPR trials are about to begin in people—But we still don't know how well it works in monkeys, *MIT Technology Review*, April 11.

38. Ma, H., Marti-Gutierrez, N., Park, S.-W., Wu, J., Lee, Y., Suzuki, K., Koski, A., Ji, D., Hayama, T., Ahmed, R., Darby, H., Van Dyken, C., Li, Y., Kang, E., Park, A. R., Kim, D., Kim, S.-T., Gong, J., Gu, Y., Xu, X., Battaglia, D., Krieg, S. A., Lee, D. M., Wu, D. H., Wolf, D. P., Heitner, S. B., Belmonte, J. C. I., Amato, P., Kim, J.-S., Kaul, S., and Mitalipov, S. (2017). Correction of a pathogenic gene mutation in human embryos, *Nature 548*, 413.

39. Quoted in Robinson, E. (2017). Study in *Nature* demonstrates method for repairing genes in human embryos that prevents inherited diseases, *OHSU News*, August 2.

40. Yong, E. (2018). The CRISPR baby scandal gets worse by the day: The alleged creation of the world's first gene-edited infants was full of technical errors and ethical blunders; here are the 15 most damning details, *The Atlantic*, December 3.

41. Cohen, J. (2018). "I feel an obligation to be balanced": Noted biologist comes to defense of gene-editing babies, *Science*, November 28.

42. Begley, S. (2019). Fertility clinics around the world asked "CRISPR babies" scientist for how-to help, *STAT*, May 28.

43. Quoted in Normille, D. (2018). CRISPR bombshell: Chinese researcher claims to have created gene-edited twins, *Science*, November 26.

44. Yong, E. The CRISPR baby scandal gets worse by the day.

45. Quoted in Reardon, S. (2019). Gene edits to "CRISPR babies" might have shortened their life expectancy: Study of almost half a million people links mutation that protects against HIV infection to an earlier death, *Nature*, June 4.

46. Wei, X., and Nielsen, R. (2019). CCR5-Δ32 is deleterious in the homozygous state in humans, *Nature Medicine 25*, 909–910.

47. Rana, P. (2018). China, unhampered by rules, races ahead in gene-editing trials, *Wall Street Journal*, January 21.

48. Prakash, A. (2019). Geopolitics of gene editing, *Genetic Literacy Project*, May 14.

49. World Health Organization. (2006). WHO gives indoor use of DDT a clean bill of health for controlling malaria, http://www.who.int/mediacentre/news/releases/2006/pr50/en/.

50. Quoted in Molteni, M. (2018). Here's the plan to end malaria with CRISPR edited mosquitoes, *Wired*, September 24.

51. The CNN interview is quoted in Livni, E. (2019). The perils and promises of redesigning animals to meet our needs, *Quartz*, May 2.

52. Quoted in Livni, E. The perils and promises of redesigning animals to meet our needs.

53. Livni, E. The perils and promises of redesigning animals to meet our needs.

54. Wegrzyn, R. (n.d.). Safe genes, DARPA, https://www.darpa.mil/program/safe-genes.

55. Nirenberg, M. W. (1967). Will society be prepared?, *Science 157*, 633.

56. Doudna, J. A., and Sternberg, S. H. *A crack in creation*.

57. Quoted in Wegrzyn, R. Safe genes.

58. Doudna, J. A., and Sternberg, S. H. *A crack in creation*.

59. McKibben, B. (2019). *Falter: Has the human game begun to play itself out?*, New York: Henry Holt.

60. Doudna, J. A., and Sternberg, S. H. *A crack in creation*.

61. Clapper, J. R. Worldwide threat assessment of the US intelligence community.

10. THIS IS NOT SCIENCE; IT IS FAKE SCIENCE

1. Otto, S. (2016). *The war on science: Who's waging it, why it matters, what we can do about it*, Minneapolis: Milkweed Editions.

2. Shermer, M. (2019). "Consilience and consensus," in *The Science Behind the Debates*, New York: Scientific American.

3. Petticrew, M. P., and Lee, K. (2011). The "father of stress" meets "big tobacco": Hans Selye and the tobacco industry, *American Journal of Public Health 101*, 411–418.

4. Oberhaus, D. (2017). The "father of stress" was a tobacco industry shill: How one of the twentieth century's most respected scientists was bought by the tobacco industry, *Motherboard—Vice*, December 7.

5. Harrett, L. (2017). Nothing but the whole (post) truth, *BioTechniques: The International Journal of Life Science Methods*, June 7.

6. Editors (2019). *The science behind the debates*, New York: Scientific American.

7. 3M. (2018). 3M state of science index, https://multimedia.3m.com/mws/media/1515295O/presentation-3m-state-of-science-index-2018-global-report-pdf.pdf.

8. Giussani, B. (2018). "Exponential," in *This idea is brilliant: Lost, overlooked, and underappreciated scientific concepts everyone should know* (ed. J. Brockman), New York: HarperPerennial.

9. LaFrance, A. (2015). Raiders of the lost Internet, *Atlantic*, October 14.

10. Higgins, K. (2019). "Post-truth: A guide for the perplexed," in *The Science Behind the Debates*, New York: Scientific American.

11. Campanile, G. (2017). What's "fake news"?, *60 Minutes*, March 27, CBS News.

12. Shearer, G. E. (2017). News use across social media platforms, Pew Research Center, September 7.

13. Iyengar, S., and Massey, D. S. (2018). Scientific communication in a post-truth society, *Proceedings of the National Academy of Sciences 116*, 7656–7661.

14. Broniatowski, D. A., Jamison, A. M., Qi, S., AlKulaib, L., Chen, T., Benton, A., Quinn, S. C., and Dredze, M. (2018). Weaponized health communication: Twitter bots and Russian trolls amplify the vaccine debate, *American Journal of Public Health 108*, 1378–1384.

15. Vosoughi, S., Roy, D., and Aral, S. (2018). The spread of true and false news online, *Science 359*, 1146.

16. Meyer, R. (2018). The grim conclusions of the largest-ever study of fake news, *The Atlantic*, March 8.

17. Arendt, H. (1972). *Crises of the republic: Lying in politics; civil disobedience; on violence; thoughts on politics and revolution*, New York: Mariner.

18. Vosoughi, S., Roy, D., and Aral, S. The spread of true and false news online; Oberhaus, D. (2018). Hundreds of researchers from Harvard, Yale and Stanford were published in fake academic journals, *Motherboard—Vice*, August 14.

19. Finkel, A. (2018). Science isn't broken, but we can do better: Here's how, *The Conversation*, April 17.

20. Sumner, P., Vivian-Griffiths, S., Boivin, J., Williams, A., Venetis, C. A., Davies, A., Ogden, J., Whelan, L., Hughes, B., Dalton, B., Boy, F., and Chambers, C. D. (2014). The association between exaggeration in health related science news and academic press releases: Retrospective observational study, *BMJ: British Medical Journal 349*, g7015.

21. Chambers, C. P. S., Boivin, J., Vivian-Griffiths, S., and Williams, A. (2014). Science and health news hype: Where does it come from?, *The Guardian* (blog), December 10, https://www.theguardian.com/science/blog/2014/dec/10/science-health-news-hype-press-releases-universities.

22. Hoffman, A. J. (2018). Why the web has challenged scientists' authority—And why they need to adapt, *The Conversation*, March 1.

23. Quoted in Groshek, J., and Bronda, S. (2016). How social media can distort and misinform when communicating science, *The Conversation*, June 30.

24. Hoffman, A. J. Why the web has challenged scientists' authority.

25. 3M. 3M state of science index.

26. Funk, C. (2017, Fall). Mixed messages about public trust in science, *Issues in Science and Technology 34*.

27. Zastrow, M. (2019). Singapore passes "fake news" law following researcher outcry—Academics say the regulation could stifle scholarly debate, *Nature News*, May 15.

11. THIS IS SCIENCE, NOT POLITICS

1. Fourier, J.-B. J. (1827). Mémoire sur les températures du globe terrestre et des espaces planétaires, *Mémoires de l'Académie Royale des Sciences de l'Institute de France VII*, 570–604.

2. Arrhenius, S. (1896). The influence of carbonic acid (carbon dioxide) in the air upon the temperature of the ground, *Philosophical Magazine and Journal of Science 41*, 237–276.

3. Zimmer, M. (2019). *Solutions for a cleaner, greener planet: Environmental chemistry*, Minneapolis, MN: Twenty-First Century Books.

4. Marshall, G. (2015). *Don't even think about it: Why our brains are wired to ignore climate change*, Bloomsbury USA; Broecker, W. S. (1975). Climatic change: Are we on the brink of a pronounced global warming?, *Science 189*, 460–463.

5. Quoted in Kahn, B. (2016). The world passes 400 ppm threshold: Permanently, Climate Central, http://www.climatecentral.org/news/world-passes-400-ppm-threshold-permanently-20738.

6. Zimmer, M. *Solutions for a cleaner, greener planet.*

7. Zimmer, M. *Solutions for a cleaner, greener planet.*

8. National Aeronautics and Space Administration—Goddard Institute for Space Studies. (2019). GISS surface temperature analysis (GISTEMP), https://data.giss.nasa.gov/gistemp/.

9. Quoted in Brulle, R. (2018). 30 years ago global warming became front-page news—and both Republicans and Democrats took it seriously, *The Conversation*, June 19.

10. Carrington, D. (2017). Record-breaking climate change pushes world into "uncharted territory," *The Guardian*, March 20, https://www.theguardian.com/environment/2017/mar/21/record-breaking-climate-change-world-uncharted-territory; NOAA, National Centers for Environmental Information. (2017). Billion-dollar weather and climate disasters: Time series, 3rd quarter release, October 6, https://www.ncdc.noaa.gov/billions/time-series.

11. Funk, C., and Kennedy, B. (2016). Public knowledge about science has a limited tie to people's beliefs about climate change and climate scientists, Pew Research Center, October 4.

12. Brulle, R. 30 years ago global warming became front-page news.

13. Quoted in Marshall, G. *Don't even think about it.*

14. Funk, C., and Rainie, L. (2015). Public and scientists' views on science and society, Pew Research Center, January 29.

15. Marshall, G. *Don't even think about it.*

16. Maibach, E., Mazzone, R., Drost, R., and Myers, T. (2017, October). TV weathercasters' views of climate change appear to be rapidly evolving, *Bulletin of the American Meteorological Society*, 2061–2064.

17. Quoted in Jazeera, A. (2018). The persistence of climate scepticism in the media, *The Listening Post*, January 28.

18. Rosling, H., Rosling, O., and Rönnlund, A. R. (2018). *Factfulness: Ten reasons we're wrong about the world—And why things are better than you think*, first ed., New York: Flatiron Books.

19. Brulle, R. 30 years ago global warming became front-page news.

20. Marshall, G. *Don't even think about it.*

21. Brulle, R. 30 years ago global warming became front-page news.

22. Quoted in Marshall, G. *Don't even think about it.*

23. Oreskes, N. (2019). "How to break the climate deadlock," in *The science behind the debates*, New York: Scientific American.

24. McKibben, B. (2019). *Falter: Has the human game begun to play itself out?*, New York: Henry Holt.

25. Raworth, K. (2017). *Doughnut economics: Seven ways to think like a 21st century economist*, White River Junction, VT: Chelsea Green Publishing.

26. Coen, D. (2018). The 19th-century tumult over climate change—and why it matters today, *The Conversation*, September 10.

27. Hayhoe, K. (2018). When facts are not enough, *Science 360*, 943.

28. Hayhoe, K. When facts are not enough.

29. Crease, R. (2019). *The workshop and the world: What ten thinkers can teach us about science and authority*, New York: W. W. Norton & Company.

30. Marvel, K. (2018). Why I won't debate science, *Scientific American Blog*, June 14, https://blogs.scientificamerican.com/hot-planet/why-i-wont-debate-science/.

31. Quoted in Marshall, G. *Don't even think about it*.

32. Quoted in Marshall, G. *Don't even think about it*.

33. Quoted in Marshall, G. *Don't even think about it*.

34. Schiermeier, Q. (2018). Droughts, heatwaves and floods: How to tell when climate change is to blame; Weather forecasters will soon provide instant assessments of global warming's influence on extreme events, *Nature 560*, 20–22.

35. Quoted in Schiermeier, Q. Droughts, heatwaves and floods.

36. Quoted in Schiermeier, Q. Droughts, heatwaves and floods.

37. McKibben, B. *Falter*.

38. Hussain, A., Ali, S., Ahmed, M., and Hussain, S. (2018). The anti-vaccination movement: A regression in modern medicine, *Cureus 10*, e2919.

39. Hussain, A., Ali, S., Ahmed, M., and Hussain, S. The anti-vaccination movement.

40. Mnookin, S. (2012). *The panic virus: The true story behind the vaccine-autism controversy*, New York: Simon and Schuster.

41. Boseley, S. (2018). How disgraced anti-vaxxer Andrew Wakefield was embraced by Trump's America, *The Guardian*, July 18.

42. Boseley, S. How disgraced anti-vaxxer Andrew Wakefield was embraced by Trump's America.

43. Broniatowski, D. A., Jamison, A. M., Qi, S., AlKulaib, L., Chen, T., Benton, A., Quinn, S. C., and Dredze, M. (2018). Weaponized health communication: Twitter bots and Russian trolls amplify the vaccine debate, *American Journal of Public Health 108*, 1378–1384.

44. Quoted in Boseley, S. How disgraced anti-vaxxer Andrew Wakefield was embraced by Trump's America.

45. Editors. *The science behind the debates*.

46. Rosling, H., Rosling, O., and Rönnlund, A. R. *Factfulness*.

47. Centers for Disease Control and Prevention. (2018). Measles cases and outbreaks: Measles 2018, https://www.cdc.gov/measles/cases-outbreaks.html.

48. Stanley-Becker, I. (2018). Anti-vaccination stronghold in N.C. hit with state's worst chickenpox outbreak in 2 decades, *Washington Post*, November19.

49. Reich, J. (2019). What's wrong with those anti-vaxxers? They're just like the rest of us, *The Conversation*, May 22.

50. Reich, J. What's wrong with those anti-vaxxers?

51. Motta, M. D. S., and Haglin, K. (2018). Countering misinformation about flu vaccine is harder than it seems, *The Conversation*, December 6.

52. Gerson, M. (2018). Are you anti-GMO? Then you're anti-science, too, *Washington Post*, May 3.

53. Kennedy, B., Hefferon, M., and Funk, C. (2018). Americans are narrowly divided over health effects of genetically modified foods, Pew Research Center, November 19.

54. South, P. F., Cavanagh, A. P., Liu, H. W., and Ort, D. R. (2019). Synthetic glycolate metabolism pathways stimulate crop growth and productivity in the field, *Science 363*, eaat9077.

55. Quoted in Benjamin, C. (2019). Scientists engineer shortcut for photosynthetic glitch, boost crop growth by 40 percent, Carl R. Woese Institute for Genomic Biology—Where Science Meets Society, January 3.

56. Shapiro, J. P. (2018). The thinking error at the root of science denial, *The Conversation*, May 8.

57. Rosling, H., Rosling, O., and Rönnlund, A. R. *Factfulness*.

58. Iyengar, S., and Massey, D. S. (2018). Scientific communication in a post-truth society, *Proceedings of the National Academy of Sciences 116*, 7656–7661.

59. Board, N. S. (2018). Science and technology: Public attitudes and understanding, *Science and Engineering Indicators 2018*, National Science Foundation.

60. Gerson, M. Are you anti-GMO?

12. QUACKERY

1. Kang, L., and Pedersen, N. (2017). *Quackery—A brief history of the worst ways to cure everything*, New York: Workman Publishing.

2. Quoted in Keslar, L. (2018). The rise of fake medical news, *Proto—The Massachusetts General Hospital*, June 18.

3. Kakkilaya, B. S. (2012). Malaria site: History of malaria treatment, https://www.malariasite.com/.

4. Hester, J. L. (2018). When quackery on the radio was a public health crisis, *Atlas Obscura*, January 12.

5. Hester, J. L. When quackery on the radio was a public health crisis.

6. Quoted in Harness, C. (2019). Healthy living: Misinformation, *Kern Valley Sun*, January 22.

7. Cole, C. A. (2017). Quincy Biosciences: What the decision means for advertising of health claims, and what it means to the FTC, November 22, https://www.retailconsumer productslaw.com/2017/11/quincy-biosciences-advertising-health-claims.

8. Quoted in Keslar, L. The rise of fake medical news.

9. Knoepfler, P. S., and Turner, L. G. (2018). The FDA and the US direct-to-consumer marketplace for stem cell interventions: A temporal analysis, *Regenerative Medicine 13*, 19–27.

10. Schwartz, L. M., and Woloshin, S. (2019). Medical marketing in the United States, 1997–2016, *Journal of the American Medical Association 321*, 80–96.

11. U.S. Food and Drug Administration. (2017). FDA warns about stem cell therapies, *FDA Consumer Updates*, https://www.fda.gov/ForConsumers/ConsumerUpdates/ucm286155 .htm.

12. U.S. Food and Drug Administration. FDA warns about stem cell therapies.

13. Da Cruz, L., Fynes, K., Georgiadis, O., Kerby, J., Luo, Y. H., Ahmado, A., Vernon, A., Daniels, J. T., Nommiste, B., Hasan, S. M., Gooljar, S. B., Carr, A.-J. F., Vugler, A., Ramsden, C. M., Bictash, M., Fenster, M., Steer, J., Harbinson, T., Wilbrey, A., Tufail, A., Feng, G., Whitlock, M., Robson, A. G., Holder, G. E., Sagoo, M. S., Loudon, P. T., Whiting, P., and Coffey, P. J. (2018). Phase 1 clinical study of an embryonic stem cell–derived retinal pigment epithelium patch in age-related macular degeneration, *Nature Biotechnology 36*, 328.

14. University of California–Santa Barbara. (2018). Stem cells treat macular degeneration—Researchers helped develop a specially engineered retinal patch to treat people with sudden, severe sight loss, *Science Daily*, March 19.

15. McGinley, L., and Wan, W. (2018). Miracle cures or modern quackery? Stem cell clinics multiply, with heartbreaking results for some patients, *Washington Post*, April 29.

16. Fu, W., Smith, C., and Turner, L. (2019). Characteristics and scope of training of clinicians participating in the US direct-to-consumer marketplace for unproven stem cell inter-ventions, *Journal of the American Medical Association 321*, 2463.

17. McGinley, L., and Wan, W. Miracle cures or modern quackery?

18. Quoted in McGinley, L., and Wan, W. Miracle cures or modern quackery?

19. Quoted in McGinley, L., and Wan, W. Miracle cures or modern quackery?

20. Clive, M., McCay, F. P., and Lunsford, W. (1956). Experimental prolongation of the life span, *Bulletin of the New York Academy of Medicine 32*, 91–101.

21. Quoted in Scudellari, M. (2015). Ageing research: Blood to blood, *Nature 517*, 426–429.

22. Villeda, S. A., Plambeck, K. E., Middeldorp, J., Castellano, J. M., Mosher, K. I., Luo, J., Smith, L. K., Bieri, G., Lin, K., Berdnik, D., Wabl, R., Udeochu, J., Wheatley, E. G., Zou, B., Simmons, D. A., Xie, X. S., Longo, F. M., and Wyss-Coray, T. (2014). Young blood reverses age-related impairments in cognitive function and synaptic plasticity in mice, *Nature Medicine 20*, 659.

23. Quoted in Scudellari, M. Ageing research.

24. Quoted in Kaiser, J. (2016). Young blood antiaging trial raises questions, *Science*, Au-gust 1, https://www.sciencemag.org/news/2016/08/young-blood-antiaging-trial-raises-ques

tions.

25. Easter, M. (2019). People are getting transfusions with young people's blood to fight aging, *Men's Health*, January 18.

26. Brodwin, E. (2018). A controversial startup that charges $8,000 to fill your veins with young blood is opening its first clinic, *Business Insider*, September 24.

27. Robbins, R. (2018). How a society gala was used to sell young-blood transfusions to baby boomers desperate to cheat death, *STAT*, March 2.

28. Quoted in Robbins, R. How a society gala was used to sell young-blood transfusions.

29. Robbins, R. How a society gala was used to sell young-blood transfusions.

30. Quoted in Keslar, L. The rise of fake medical news.

31. Korownyk, C., Kolber, M. R., McCormack, J., Lam, V., Overbo, K., Cotton, C., Finley, C., Turgeon, R. D., Garrison, S., Lindblad, A. J., Banh, H. L., Campbell-Scherer, D., Vandermeer, B., and Allan, G. M. (2014). Televised medical talk shows—What they recommend and the evidence to support their recommendations: A prospective observational study, *BMJ: British Medical Journal 349*, g7346.

32. Inoue-Choi, M., Oppeneer, S. J., and Robien, K. (2013). Reality check: There is no such thing as a miracle food, *Nutrition and Cancer 65*, 165–168.

33. Briggs, B. (2015). Dr. Oz responds to critics: "It's not a medical show," *NBC News Health Report*, April 23; Belluz, J., and Hoffman, S. (2015). All of the arguments Dr. Oz made against his critics were wrong, *Vox*, April 23.

34. Institute, G. W. (2018). Wellness now a $4.2 trillion global industry—with 12.8% growth from 2015–2017, *2018 Global Wellness Economy Monitor*, https://globalwellness institute.org/press-room/press-releases/wellness-now-a-4-2-trillion-global-industry/.

35. Singh Ospina, N., Phillips, K. A., Rodríguez-Gutierrez, R., Castaneda-Guarderas, A., Gionfriddo, M. R., Branda, M. E., and Montori, V. M. (2019). Eliciting the patient's agenda—secondary analysis of recorded clinical encounters, *Journal of General Internal Medicine 34*, 36–40.

36. Quoted in Biba, E. (2019). Why are so many women rejecting medical science? *DAME*, April 22, https://www.damemagazine.com/2019/2004/2022/why-are-so-many-women-reject-ing-
medical-science/.

37. Quoted in Biba, E. Why are so many women rejecting medical science?

38. Quoted in Biba, E. Why are so many women rejecting medical science?

39. Rothstein, C. (2019). The wellness industry isn't making you well: It's exclusionary, expensive, and too trendy for its own good, *Marie Claire*, January 1.

40. Gunter, J. (2018). Worshiping the false idols of wellness, *New York Times*, August 1.

41. Suazo, A. (n.d.). 10 activated charcoal recipes to detox from the inside out, https://blog.bulletproof.com/activated-charcoal-recipes-2b3c4t5c/.

42. U.S. Food & Drug Administration. (n.d.). Health fraud scams, https://www.fda.gov/ForConsumers/ProtectYourself/HealthFraud/default.htm.

43. Federal Trade Commission. (n.d.). Consumer information—Health & fitness, https://www.consumer.ftc.gov/health.

13. SOME THOUGHTS ON KEEPING SCIENCE HEALTHY IN A SUSTAINABLE FUTURE

1. Johnson, E. (2019). The decline of trust in science "terrifies" former MIT president Susan Hockfield, *Vox—Recode*, May 31.

2. McKibben, B. (2019). *Falter: Has the human game begun to play itself out?*, New York: Henry Holt.

3. Persson, I., and Savulescu, J. (2014). *Unfit for the future: The need for moral enhancement*, Oxford: Oxford University Press.

4. Rozenblit, L., and Keil, F. (2002). The misunderstood limits of folk science: An illusion of explanatory depth, *Cognitive Science 26*, 521–562.

5. Goldman, G. T. (2019). Trump's plan would make government stupid: Cuts to science advisory panels for federal agencies will haunt the United States long after the current administration finishes, says Gretchen T. Goldman, *Nature 570*, 417.

6. Masood, E. (2018). Why academic freedom is needed more than ever, *Nature 563*, 620.

7. Sullivan, M., Sellers, C., Fredrickson, L., and Lamdan, S. (2019). The EPA has backed off enforcement under Trump—Here are the numbers, *The Conversation*, January 3.

8. Piller, C. (2019). Exclusive: FDA enforcement actions plummet under Trump, *Science*, July 2, https://www.sciencemag.org/news/2019/2007/exclusive-fda-enforcement-actions-plummet-under-trump.

9. Crease, R. (2019). *The Workshop and the world: What ten thinkers can teach us about science and authority*, New York: W. W. Norton & Company.

10. Davenport, C., and Landler, M. (2019). Trump administration hardens its attack on climate science, *New York Times*, May 27, https://www.nytimes.com/2019/2005/2027/us/politics/trump-climate-science.html.

11. Bump, P. (2019). Trump claims that wind farms cause cancer for very Trumpian reasons, *Washington Post*, April 3, https://www.washingtonpost.com/politics/2019/04/03/trump-claims-that-wind-farms-cause-cancer-very-trumpian-reasons/?utm_term=.681851da89 f0.

12. Quoted in Flavelle, C. (2018). Academics need not apply: Trump's agencies cool to professors, *Bloomberg*, March 12, https://www.bloomberg.com/news/articles/2018-2003-2012/academics-need-not-apply-trump-s-agencies-cool-to-professors.

13. Johnson, E. The decline of trust in science "terrifies" former MIT president.

14. Tollefson, J. (2018). China declared world's largest producer of scientific articles, *Nature 553*, 390.

15. Lander, E. S. (2018). Will America yield its position as the world's leader in science and technology? *Boston Globe*, January 29.

16. Fleming, L., Greene, H., Li, G., Marx, M., and Yao, D. (2019). Government-funded research increasingly fuels innovation, *Science 364*, 1139.

17. Quoted in Johnson, E., The decline of trust in science "terrifies" former MIT president Susan Hockfield.

18. McCarthy, N. (2018). International student enrollment declining in the U.S., *Statista.com*, May 31, https://www.statista.com/chart/14052/international-student-enrollment-declining-in-the-us/.

19. Cheng, Y. (2019). My science has no nationality, *SupChina*, May 29, https://supchina.com/2019/2005/2029/my-science-has-no-nationality/.

20. Tollefson, J. (2019). Chinese American scientists uneasy amid crackdown on foreign influence, *Nature 570*, 13–14.

21. Reif, L. R. (2019). Letter to the MIT community: Immigration is a kind of oxygen, *MIT News*, June 25, http://news.mit.edu/2019/letter-community-immigration-is-oxygen-0625.

22. Baker, K. (2019). Trump's immigration plan would have missed this Nobel Prize winner: I. I. Rabi came up with the idea behind M.R.I.s; who brought it to fruition? A man whose family fled genocide, *New York Times*, June 24.

23. Saffo, P. (2018). "Haldane's rule of the right size," in *This idea is brilliant: Lost, overlooked, and underappreciated scientific concepts everyone should know* (ed. J. Brockman), New York: HarperPerennial.

Index